HUMAN FACTORS AND VOICE INTERACTIVE SYSTEMS

THE KLUWER INTERNATIONAL SERIES
IN ENGINEERING AND COMPUTER SCIENCE

HUMAN FACTORS AND VOICE INTERACTIVE SYSTEMS

edited *by*

Daryle Gardner-Bonneau

Office of Research, Michigan State University
Kalamazoo Center for Medical Studies
Kalamazoo, Michigan

Kluwer Academic Publishers
Boston/Dordrecht/London

Distributors for North, Central and South America:
Kluwer Academic Publishers
101 Philip Drive
Assinippi Park
Norwell, Massachusetts 02061 USA
Tel: 781-871-6600
Fax: 781-871-6528
E-mail: kluwer@wkap.com

Distributors for all other countries:
Kluwer Academic Publishers Group
Distribution Centre
Post Office Box 322
3300 AH Dordrecht, THE NETHERLANDS
Tel: 31 78 6392 392
Fax: 31 78 6546 474
E-mail: orderdept@wkap.nl

Electronic Services: http://www.wkap.nl

Library of Congress Cataloging-in-Publication Data

A C.I.P. Catalogue record for this book is available
from the Library of Congress.

Printed on acid-free paper.

Printed in the United States of America

This book is dedicated to the memory of
Dr. Gary K. Poock - pioneer, mentor,
colleague, and friend

Table of Contents

List of Figures

List of Tables

LIST OF CONTRIBUTORS

Bruce Balentine
Same Page Design Group

Sara Basson
IBM T.J. Watson Research Center

Harry E. Blanchard
AT&T Labs

Susan J. Boyce
AT&T

Catalina Danis
IBM T. J. Watson Research Center

Michel Divay
Université de Rennes
Institut Universitaire de Technologie
France

Alexander L. Francis
Department of Psychology
The University of Chicago

Daryle Gardner-Bonneau
Michigan State University
Kalamazoo Center for Medical Studies

Mária Gósy
Phonetics Laboratory,
Research Institute for Linguistics,
Hungarian Academy of Sciences

Brian Hansen
Oregon Heath Sciences University

John Karat
IBM T. J. Watson Research Center

Jennifer Lai
IBM T. J. Watson Research Center

Steven H. Lewis
AT&T Labs

Martha J. Lindeman
Users First, Inc.

Arnold M. Lund
U S WEST Advanced Technologies

Catherine R. Marshall
CollabTech, Inc.

Géza Németh
Dept. of Telecommunications and
Telematics
Technical University of Budapest
Hungary

David G. Novick
EURISCO, and
Oregon Graduate Institute

Howard C. Nusbaum
Department of Psychology
The University of Chicago

Gábor Olaszy
Phonetics Laboratory,
Research Institute for Linguistics,
Hungarian Academy of Sciences

Stephen Sutton
Fluent Speech Technologies, Inc.

John C. Thomas
IBM T.J. Watson Research Center

Catherine Wolf
IBM T. J. Watson Research Center

PREFACE

A number of books have been written about speech technologies. Most of them are devoted, specifically, to the technologies under development, rather than the application of these technologies in the real world. It has been only recently that people have started writing books about the technology in use (e.g., Baber and Noyes, 1993; Markowitz, 1996; Raman, 1997; Schmandt, 1994; Syrdal, Bennett. and Greenspan (1995). All of them consider human factors engineering to a greater or lesser degree but, curiously, some (e.g., Schmandt, 1994; Raman, 1997) never emphasize the vital role that human factors has played, and will continue to play, in the design of successful applications employing these technologies.

Human factors engineering has a long history dating back, at least, to the industrial efficiency studies of the early part of this century (see Sanders and McCormick (1993) for a discussion of the profession's history). It is the discipline that strives to ensure that technologies are used in a way that enhances human productivity and task performance, minimizes errors, and ensures safety. As a discipline, it has had an impact in virtually every area of human endeavor. From the design of aircraft cockpits and nuclear power plant control rooms, to the design of easy-grip gardening tools and the toothbrush in your bathroom, the application of human factors engineering has helped to ensure that products and systems are easy to use, effective, and safe. It does this through methods, tools, and techniques that focus on the user, or the potential user, from the very beginning of the design process through to the finished product. Its principal tenet is that, to be successful, applications of technologies must be designed with due consideration of the user; applications that neglect the user interface are destined to fail.

One of the interesting, and frustrating, things about this field is that people acknowledge its contribution, primarily, when a product, piece of equipment, or system under discussion is lacking good human factors engineering. They lament that it is unusable, difficult to learn, prone to errors, and unable to meet their expectations and fulfill their goals. Alternatively, when products are successful, good human factors engineering is rarely credited with the success. This is one of the reasons I wanted to write this book – to showcase what it is that human factors professionals do, behind the scenes, that contributes to successful products.

Another characteristic of the human factors field is the fact that its literature is scattered everywhere. Because human factors engineering applies to the design of nearly everything, papers get presented and published in many different forums.

Papers about telecommunications get published in telecommunications journals, papers about medical systems in medical journals, and papers about speech technologies get presented and published through the various forums devoted to these technologies. Human factors engineering does have its own national organization in the United States (The Human Factors and Ergonomics Society), which publishes papers on human factors engineering in all areas to which it is applied and sponsors an annual conference that addresses human factors across all domains of application. However, many technology developers in individual application domains may never encounter these journal and conference papers or this organization. To complicate matters further, for every human factors professional whose primary professional affiliation is The Human Factors and Ergonomics Society, there is another who is more involved with the Association for Computing Machinery (ACM) and, specifically, its Special Interest Group on Computer-Human Interaction (SIG CHI). Therefore, many professionals in the human factors field may not encounter each other on a regular basis, because they attend different professional meetings. Thus, another reason I wanted to write this book was to bring much of the human factors literature related to speech technology together in one place.

Finally, many of the books written on speech technology applications emphasize the desktop environment. I am not aware of any that focus on over-the-telephone use of speech technologies, even though such applications are among the most ubiquitous, at least in the United States. There is an especially rich history of human factors work in telephony, dating back to Bell Laboratories at the turn of the century, and much of the human factors contribution to speech technology applications has been in the area of interactive voice response (IVR) systems. Nearly half of this book is devoted to human factors work on speech-enabled telephony applications, both those with touch-tone interfaces and those under development more recently that employ speech recognition front-ends.

I had hoped to write my own book about human factors and speech technologies. However, things being what they are in the life of a working mother these days, it just wasn't in the cards. But rather than abandon my goals, I decided to pursue them with the assistance of some of my distinguished colleagues.

The resulting book is divided, roughly, into three sections devoted to 1) automatic speech recognition; 2) synthetic speech; and 3) interactive voice response. Some of the chapters are devoted to specific applications of the speech technology, while others are issue-oriented or provide a comprehensive view of the work in a specific applications area. Together, they give a fairly comprehensive picture of the kinds of work human factors professionals do, and the ways in which they accomplish it.

The first two chapters are devoted to applications of automatic speech recognition (ASR) and a discussion of work ongoing in the area of natural language processing. In the first chapter, John Karat and his colleagues at IBM discuss, in detail, the design work they have done on the user interfaces of several products. One of these products in particular – MedSpeak – is enjoying wide acceptance in the medical transcription marketplace. IBM's attention to the needs of users, described in detail for all of the applications discussed, serves to highlight the importance of the up-front work on needs assessment that can be a major determinant of an application's success.

Natural language processing is the logical next step for speech technology, but its application lies somewhere in the future. In Chapter Two, Susan Boyce, of AT&T, demonstrates with copious examples just why this is the case. At the same time, she highlights the progress that has been made in the last several years. Understanding communication at the level of conversations is required for natural language processing, and it is a very difficult task, indeed.

The next three chapters, all very different, are devoted to synthetic speech. The first, by Alexander Francis and Howard Nusbaum, provides a comprehensive view of the factors that determine the quality of synthetic speech – namely, intelligibility and naturalness - and the tools and techniques used in its evaluation. The following two chapters give the reader a glimpse at some of the very difficult work that makes synthetic speech applications possible. In Chapter 4, Michel Divay describes the challenging and tedious effort involved in doing grapheme-to-phoneme transcription, both for English and for French. The fact that such transcription is language-dependent serves to illustrate just how much work is involved to convert text to speech in any language, and how daunting the task may be to achieve this result for the many languages of the world that have not been synthesized to date.

It has long been my belief that one of the areas in which speech technologies could best serve humanity is in the service of people with disabilities. Although some well known applications exist, and an innovative paper or two appears occasionally, we have only scratched the surface of what is possible with respect to assistance for people with visual, speech, hearing, language, and mobility impairments. One of the areas in which speech technologies have had little application, but where much is possible, is in the assessment of speech, language, and hearing disorders and therapies for these disorders. Thus, it is a special pleasure to include Maria Gósy's contribution (Chapter 5) on the use of synthesized speech for evaluating children's hearing and acoustic-phonetic perception. Chapter 6 continues with this theme, and presents material on the uses of both synthetic speech and speech recognition in assistive technology, and to achieve the goal of universal access – the design of systems to be usable by everyone, regardless of his or her level of disability.

Seven of the remaining eight chapters are devoted to telephony-based applications of speech technology. In the first of these (Chapter 7), I describe efforts to develop user interface guidelines for interactive voice response (IVR) applications, and how U.S. approaches to IVR design, predicated on the availability of touch-tone technology, need to change now that speech recognition has "come of age." In the creative chapter that follows, David Novick and his colleagues discuss a task-based approach for determining whether development of an IVR application is appropriate, given the particular features of the task situation involved.

When the prospectus for this book was evaluated, several reviewers questioned the inclusion of Arnold Lund's contribution (Chapter 9), because it does not address the use of speech technology, specifically. It has remained, however, for two reasons. First, because touch-tone-based applications for IVR systems are predominant in the United States, and will be for some time to come, Lund's analysis of touch-tone dialing errors will continue to be relevant to the design of these applications. Second, this chapter exemplifies the quantitative, empirical nature of much of the work human factors specialists do. For the many speech scientists and electrical engineers out there who tend to consider human factors work as "soft and squishy" and unquantifiable, this contribution, which is typical of those found in human factors journals, demonstrates that there is as much science as art in this field.

Chapter 10, by Bruce Balentine, offers a view that appears contradictory to the one I present in the first IVR chapter, at least at first glance. However, this is not the case, really. While I discuss how IVR applications can and should change as a result of the introduction of automatic speech recognition, Bruce shows us how we can live within the framework of the menu-driven design approach we've created in the U.S., whether an application is speech-enabled or not. Endlessly creative, Bruce shows us how to make menu-based designs work, by careful attention to the construction of the IVR dialogue.

The second of two chapters from the Research Institute for Linguistics in Budapest, Hungary (the first was Gósy's chapter), Chapter 11 describes the process used to develop high quality concatenated speech for a banking-related IVR application. The authors, Gábor Olaszy and Géza Németh, also describe their evaluation study of a resulting application, and discuss both the perception and the acceptability of the synthesized speech.

Chapter 12, by Harry Blanchard and Steven Lewis, provides a comprehensive treatment of voice mail applications. Detailed design guidelines are presented, and the authors discuss the ways in which voice mail is now being integrated with e-mail and FAX services to achieve unified messaging. Finally, Martha Lindeman describes in Chapter 13 an approach to computer-telephony integration (CTI) problems through a design framework that considers the dimensions of time and complexity in a call's life cycle.

In the concluding chapter, I provide some final thoughts about applications that received limited attention in the book and offer some musings about the future of this technology and the continuing role of human factors in speech-enabled application design. It is my hope that speech application developers will derive benefit from the many "lessons learned" presented in this book, and will make use of both human factors expertise and the techniques and tools of this field as they design tomorrow's applications.

Daryle Gardner-Bonneau

REFERENCES

Baber, C., and Noyes, J. M., Eds. (1993). *Interactive speech technology: Human factors issues in the application of speech input/output to computers.* London: Taylor & Francis.

Markowitz, J. A. (1996). *Using speech recognition.* Upper Saddle River, NJ: Prentice-Hall.

Raman, T. V. (1997). *Auditory user interfaces: Toward the speaking computer*. Boston: Kluwer Academic Publishers.

Sanders, M., and McCormick, E. J. (1993). *Human factors in engineering and design* (7th ed.). New York: McGraw-Hill.

Schmandt, C. (1994). *Voice communication with computers: Conversational systems*. New York: Van Nostrand Reinhold.

Syrdal, A., Bennett, R., and Greenspan, S., Eds. (1995). *Applied speech technology*. Boca Raton: CRC Press.

ACKNOWLEDGEMENTS

In my professional career, I've edited individual articles, special issues of journals, and even a magazine, but never a book. Thus, I must thank, first, the contributors, who trusted me with their manucripts and cooperated marvelously during the editing process, maintaining their support, good humor, and high spirits throughout. Some are friends; others I'd never met and know only by reputation. But I'm proud to call all of them my colleagues. I also thank Alex Greene at Kluwer, who backed this project and provided gentle nudges, as needed. It would have been impossible to complete this project, however, without the love and support of those dearest to me. In this regard, I thank my son, Nicholas, for enduring so many evenings of fast food, and for tolerance, well beyond his years, during those days when I had so little quality time to spend with him. Last, but only on this page, I thank John Bonneau – husband extraordinaire – for all the time and effort he expended to produce the camera-ready version of the book. More importantly, I thank him for giving me, through his endless love and support, the courage to pursue all my dreams and the confidence to capture so many of them.

1 SPEECH USER INTERFACE EVOLUTION

John Karat, Jennifer Lai, Catalina Danis and Catherine Wolf

IBM T. J. Watson Research Center

Key words: Automatic Speech Recognition (ASR), Dictation, StoryWriter, MedSpeak, Conversation Machine for Banking

Abstract: *In some naive past, some of us assumed that the introduction of speech recognition and output to interactive systems would undoubtedly improve their usability. We now know that designing computer systems that accept speech as input or use speech as output requires considerable effort. While speech may be a natural form of communication for humans, speaking to computer artifacts or having them speak to us is a different activity. We are only beginning to become aware of the subtle ways in which such differences manifest themselves as we attempt to make use of speech technologies that, while they are vastly improved over research prototypes of 10 years ago, still have gaps in performance compared to what we can expect of human-human communication. We present some of the lessons we have learned over the past years as we have been involved in the design of systems that make use of speech technology. While the principles of feedback to the user are the same for different applications of speech technology, the particular design decisions depend on the task, user, and context of use.*

1. INTRODUCTION

Automatic speech recognition (ASR) technology has been under development for over 25 years, with considerable industry and government resources being devoted to developing systems which can translate speech input into character strings or commands. After all of this effort we are just beginning to see fairly wide application of the technology. It might be possible to interpret this relatively slow penetration of ASR into interfaces for computer systems as an indication that speech is not a good modality for such interfaces, and that efforts to develop ASR are naive in some way. After all, look at how quickly the Internet has spread. While this "good ideas catch on quickly" notion has a lot of appeal, it would be incorrect to use it at this time in considering ASR. Simply put, ASR will be widely used in interacting with computers in the not to distant future. It will not necessarily replace other input modalities, but it will be a very powerful means of human-computer communication.

However, we think that there are some fundamental factors to keep in mind when considering the value of ASR and how rapidly and widely it will spread. First is the **ongoing development of speech recognition and generation technology** itself. How we go from an acoustic signal to some computationally useful translation of the signal remains technically challenging. Additionally, speech recognition is of course only part of the picture - a complete speech user interface (SUI) might be expected to produce speech as well. We are in a time of significant changes and improvements in our ability to recognize free-form speech and to generate natural sounding speech output. The goals are speaker-independent, continuous-speech recognition and human sounding speech output. These are not so far off as they once were, but we are not there yet.

Second, while we like to think that speech is a natural form of communication, it is misleading to think that this means that it is easy to build interfaces that will provide a natural interaction with a non-human machine. **Speech is not natural for communicating with machines**. The way people use speech in communicating is strongly influenced by their experience in using it with other people. How we might use speech in communicating with machines needs to be considered - although having no difference between human-human and human-computer communication might be a good goal, it is not one likely to be attainable in the near future. The naturalness of speech communication should be seen as reflecting our experiences in using the modality to communicate with other humans. These people share a great deal of knowledge with the speaker - something that cannot be said of current computers (though there are ongoing efforts to provide machines with broad contextual and social knowledge). A great deal of the naturalness that we take for granted in verbal communication goes away when the listener doesn't understand the meaning of what we say.

Finally we argue that **it takes time and practice to develop a new form of interaction**. The systems described in this chapter all involve efforts to develop systems that utilize speech as a part of the human-computer interaction. You can consider these as "practice" in designing to incorporate this new technology into our systems. The systems were targeted to meet needs of different groups, and also had different business objectives to attempt to fufill. For each of these systems, Personal Dictation System, StoryWriter, MedSpeak, and Conversation Machine, we will describe the design context (the problem we were trying to address and the constraints under which we operated). It is our hope that this can contribute both to an understanding of how to incorporate speech into other systems and, more generally, how to incorporate new technology into useful systems.

There are some common problems that we faced in all of the systems described below. The three characteristics of speech and speech recognition technology that presented us with significant challenges were recognition errors, the lag separating spoken word and its transcription, and the difficulty of knowing what parts of the system particular speech strings were directed to. Dealing with errors is different in speech recognition (where someone can say something they intend to say but the system recognizes something different as being said) than with other input devices where "human error" is the main factor to be addressed (Danis and Karat, 1995). Dealing with recognition lag is partly related to processing requirements and partly related to recognition algorithms that make use of following context. Dealing with context (or mode) issues can involve word or command selection and differentiating this activity from dictation of text. For each of the systems we will have something to say about how these issues were addressed.

This chapter contains four case studies in the design of systems which make use of speech technology. Any such design case study could go into the details of hundreds of design decisions at great lengths, but our effort here is to focus our discussions on some of the issues we felt were particularly salient because of the role of speech technology. The systems represent developments during the 90's as work at our research center evolved from a focus on recognition of isolated words to continuous speech and from speech as input only to speech for both input and output. Table 1 summarizes the four systems that we will discuss in this chapter.

Table 1.   General characteristics of speech recognition systems considered here.

System	General Characteristics	Design Challenge
Personal Dictation System	An early (1993) general purpose, large vocabulary desktop dictation system.	Bringing discrete speech recognition to general users.
StoryWriter	An early (1993) ASR system targeted to journalists with RSI.	Mimimizing the need for non-speech interactions.
MedSpeak	A continuous speech recognition system targeted for Radiologists.	Continuous recognition in a demanding environment.
Conversation Machine	A system for speech conversations targeted to banking transactions.	Speech as input and output integrated with a visual display.

2.   PERSONAL DICTATION SYSTEM

This section describes the development of a large vocabulary speech recognition system intended for general audiences using personal computers. IBM had engaged in research in speech recognition since the 1970s, but had not attempted to market a general-purpose speech recognition product based on its research technologies until the 1990s. The work of the group had long focused on developing algorithms that were fast and reliable enough to approximate human capabilities. While demonstrations of the technology had taken place over the long period of work, widespread application did not seem possible until recently. The first "shrink-wrapped" product (one intended for a large general market) making use of the technology developed by the group was called the IBM Personal Dictation System (IPDS) - the first offering in a product series now called ViaVoice. Released at the beginning of 1994, IPDS allowed the user to dictate using isolated word speech (that is, with a short pause between words) and to control some of the functions of personal computer desktops. The system required the users to start with a training session in which about 200 sentences were read (requiring about one hour to complete). This information was used to develop a speaker specific voice model that the system used in matching speech input. While error rates are very difficult to cite with certainty, it was common for a user's initial performance in using the system to dictate text to result in accuracy in the low- to mid-90% range for "in vocabulary" words (i.e., words that were in the approximately 20,000 word vocabulary shipped with the system). Experience with the system reduces individual error rates, as words are added to the known vocabulary, as the system learns individual pronunciations, and as practice dictating increases. However, even with frequent use an error rate of about 2-5% will remain.

IPDS was intended as a general-purpose speech-to-text system. In particular, the product was the result of a technology transfer project in which a small team was given the mission of developing a product that made use of IBM's speech recognition technology "within one year" of project inception. As such we were more interested in fielding something to serve as an entry into the marketplace than we were with meeting a specific, well-defined set of user requirements. Once the team decided on the direction for the product (other options such as developing a speech desktop navigator were also considered), sessions were conducted with potential user groups to try and understand the match between the technology capabilities and user requirements (Karat, 1995).

2.1   Users/Task/Context of Use

User Characteristics.   Since it was intended as a general-purpose system, the target audience was viewed as "anybody." This is a difficult audience to design for in that one ends up having to make very general assumptions about the capabilities of the user population. Our only real assumption was that the users would generally be familiar with word processing on a personal computer. We did not assume familiarity with any particular product, but did assume that users would not be experiencing computer text creation for the first time with IPDS.

Tasks.   What kind of requirements does a general audience have for text creation? It was clear to us that they are quite extensive and quite complex. The competing systems were viewed as the leading word processing systems of the time (e.g., Microsoft Word, Word Perfect, or AmiPro by Lotus). From the task perspective, our (obvious) first choice would have been to speech-enable an existing word processing application. While this was considered, various non-technical (i.e., business) factors prevented our taking this path at the time. We did not think that it would be possible to develop a new full function editor to address the wide audience, and instead elected to focus on providing a text entry mechanism that could be used in conjunction with all existing packages. Rather than developing our own editor, we elected to provide a "speech-window" from which text could be copied and pasted into other desktop applications through existing operating system cut-and-paste buffers. The speech window was developed from a basic speech window - a window which simply displayed the text that was recognized when someone was dictating - that was an existing part of the speech recognition algorithm group's development environment. Speech-enabling an existing word processing application was a project option that would have required considerably more outside negotiation than we were interested in pursuing at the time. (Industry-wide efforts to provide a general-purpose speech-enabling framework for word processing systems started after this project was completed.)

The speech recognition technology of early 1993 was just beginning to reach the point where developing a general-purpose tool for speech to text was considered practical. Ongoing work in the field had identified mid-90's as the minimum accuracy level necessary for user acceptance, and this meant that isolated speech and training (both viewed as drawbacks to acceptance) would be required for wide acceptance of such a product. Still, the decision was made that this was a technology domain in which IBM would want to be active, and that even with a relatively small initial market for such a product we should go forward.

Context Of Use. The general direction was to focus the design on production of medium to large quantities of text, and to view speech to text as augmentation (rather than replacement) for other input devices in a system (such as keyboard and mouse). While we sought to provide a "speech command" for carrying out all system actions, we realized early on that designing a usable "speech-only" system was beyond the resource/time constraints placed on the project. Some items, such as indicating a specific screen location, or creating objects such as tables, were simply going to take more resources than we had available (about ten person-years were allocated to the project).

2.2 Capabilities and Constraints

2.2.1 Recognition Errors

Speech recognition systems will probably always have to deal with recognition errors - errors in which one word was said (possibly clearly) but another was judged as the best match by the system. It is also a factor in speech recognition that the user is not aware of a recognition error as soon as it occurs but rather at some delay from when the word was uttered. The IPDS system (and speech recognition systems in general) take longer than the 70 ms between action and visual feedback that have generally been found to be the maximum time between events for them to be perceived as co-occurring. The delay in timing between speech event and feedback makes it impossible for users to simultaneously carry out the task of "thinking about what to say" and the task of "checking to see if the system correctly recognized what I said".

Given the reality of such lags, our experience with the dictation process for speech to text suggested that most users would find it better to make all error corrections in a single pass after completing a dictation than to make corrections "as they go." This led us to focus on supporting an error correction process which would take place well after the dictation. One way in which we tried to support this was to provide information on what was said (since users might not remember).

The resulting design allowed users to select any word in the recognized text and have the system play back the dictated word in the user's voice.

In keyboard text entry, corrections are commonly made by asking the system to check for likely errors. We thought it would be desirable to do this for speech recognition also. One thing that we considered as a possible aid to the proofreading process was to visually flag words that the engine was "less sure of." In recognizing any word, the system displays the highest probability word match. It also has information about the probability of other words matching what was said. We had speculated that it might help the correction process for users if they could easily see words that the system was "less certain of" (i.e., those falling below a certain level of probability). This did not turn out to be a useful aid to the error correction process. The level of certainty as measured by the probability measure from the recognizer did not turn out to be a sufficiently good indicator of recognition error to be of value to the user. We included a control in the initial prototypes in which the user could set a probability level for highlighting less-certain words, but it proved of little practical use.

Even though the user must to the work of determining what is in error, the main error correction mechanism was modeled after spelling correction dialogs. If a user identifies a word as incorrect, a dialog appears with a list of possible alternative words based on the probability metric from the recognition engine. Unfortunately, words that "sound similar" to the recognizer (that is, those that are determined to be likely candidates on a list of alternative words) are much less likely to be the intended word than in the analogous case of spelling checkers, in which the (correct) alternative is almost always on the provided list. In general, error correction requires users to type in the intended word, rather than their being able to select it from a list of alternatives with either speech or by pointing. In general we felt that we had provided a good error correction mechanism, but that better solutions would eventually be found.

2.2.2 Lag in Recognizing Text

When someone is entering text using a keyboard, they can reliably look at the display to see exactly what they have typed at any given moment and see a display that is current (i.e., does not lag behind that they have typed). Furthermore, when you look at a display of typed text, you are certain that text already entered will not change unless you take a specific action to change it. In addition to the simple response lag mentioned above, we faced a problem brought about by the nature of the recognition algorithms we were using. IPDS (and the speech recognition engine on which it is built) makes use of language models to assist in speech recognition. These models include statistical description of the likelihood of word sequences in the particular language and use context. For IPDS the models were based on analysis of general business correspondence. Such models adjust the likelihood judging matches of speech input to stored word pronunciations based on word trigrams, that is, the probability of a word match can be influenced by both the word preceding and the word following a given word. In using words following a given word as a part of a recognition algorithm, no matter how fast the matching algorithms were, IPDS would not make a final decision on any given word until it had input for the next word (or until an end of dictation was indicated). Thus, our speech recognition algorithm is such that what the system thinks is the best match for one word might change as additional words are dictated. We had either to compound the lag problem by suppressing display of a word until the next word had been recognized or present the user with a display that might change one word for another as additional words were spoken.

While we struggled to find a design solution that would match current keyboard to display experiences, we realized that such a design would not be possible within current technology. Therefore, we focused on understanding the impact of a changed dictation process in which watching the screen during dictation was to be discouraged or viewed as possibly distracting. One of the things we focused on in our early evaluations was how much the users would be distracted by the display. This turned out to be less of a problem than we might have feared. While people are quite used to monitoring the progress of what they are typing on the screen, it is not the case that they monitor on a keystroke by keystroke basis in which feedback from each keystroke is visually checked before they make another keystroke. If normal text entry proceeded in a process in which a unit was entered and then verified before the next unit could be entered, the IPDS display would be totally unsatisfactory. While we did not study text creation thoroughly, it seemed that for most text creation tasks general users mixed modes of creation in which there was little checking of the display with pauses during which they might check previous text or during which they thought about what to say next without referring, necessarily, to what was on the display. In either of these cases (unless the pause to check the recognized text is very short), we found that the lag in the display did not cause much difficulty with the primary text creation task.

What we did not accommodate as well as we might like are the attempts to correct recently dictated words without leaving the dictation mode. Various attempts at adding special commands to the dictation vocabulary (like "whoops" to erase the immediately preceding word) were tried, but we did not come up with a set we felt would be worthwhile adding to the system. The range of things that we wanted to support seemed just beyond the level of complexity that we felt we could reasonably accommodate. It included indicating (thorough voice) arbitrary words within the current paragraph to be changed, and also seemed to require allowing for tinkering changes and not just replacing misrecognitions since the users might not differentiate strongly between these two types of changes.

2.2.3   Mode Switching

With IPDS, speech input is provided to more than just an isolated dictation application. Voice input can be directed to any application running on a system. It is received by IPDS and appropriate text strings are passed to the application which has "speech focus." On one hand this is flexible, but on the other it creates a need to know what application is expecting input. It was not possible to have a single vocabulary for all possible applications (e.g., for Microsoft Windows, for an e-mail application, and for the dictation application); we had to switch between possible vocabularies depending on which was to receive the input. This proved fairly easy for the system, but fairly difficult for users to understand and track.

While language models and word trigrams were used in the speech recognition algorithms of IPDS for accurate large vocabulary recognition, faster recognition is possible if the expected vocabulary is much smaller. For IPDS, when the expected input was a desktop command, recognition was based on a smaller expected vocabulary (just the commands that would be legal in the current system state), and no language model information was used to augment the acoustic match. That is, there was no equivalent of the word trigram data collected or used to augment the matching process. Actually, we did not start off by assuming that such information would not be useful; we began with an understanding that unless we knew reliably whether the current input was text or command, we would get into difficult situations with recognition errors. This drove us to consider explicit indication of mode as a requirement. This decided, the next step was to ask whether the language model would be necessary for commands. (We did not have one, and hoped that sufficiently good recognition could be achieved without one.) For IPDS, user feedback has suggested that recognition accuracy in command mode without a language model is quite adequate. Unfortunately, this does not mean that the implementation of the modes, or the adequacy of the command vocabulary were completely satisfactory.

The problem that gives rise to the mode split is that command words (the word or words spoken to issue a command) can also be words that might be dictated. Before one acts on a string as a command (e.g., "delete"), one needs to be much more certain of the recognition than one needs to be for dictated text. Much of this need for certainty exists because correcting a misrecognized word in text is relatively easy compared to undoing the execution of some commands.

On one hand we felt we had a fairly clever solution to this difficult problem. We were able to inquire about the state of the desktop from the operating system and create on the fly a list of "valid commands" (e.g., the list of objects and menu bar items on the active window that might be selected). Without requiring us to build in lists of command names for all possible application programs, we could simply read assumed command names from menu bars or object names. These were names that we could use to build an active vocabulary. Unfortunately, such internal names (particularly menu bar items) were not always known or obvious to users who might have selected them for years without knowing the full object name. For example, options such as "copy" are sometimes given the display label "copy to clipboard" in application programs, and it was the full name that would end up in our vocabulary. Also, while we were selecting acceptable commands from a knowledge of the system state, which included knowing which of several windows might be currently active, users would often direct commands to other visible system objects. In a very natural process, they would look at an object and direct a speech command to it, the problem being that the system would not be aware of the information from the users' visual focus.

We never reached a completely satisfactory resolution to the "what can I say" problem. The fact that the system would recognize different words depending on the system state was never clear to users, even though we designed a number of mechanisms for highlighting the system component that was "listening" (and thus determining the active vocabulary) at the given moment. That this context switching remains tricky in speech recognition systems should not be surprising, since it can be a problem in human-human speech recognition as well. If you don't have the right context, you can be quite confused about what someone is saying, and can be unsure what the source of the confusion is.

2.3 Lessons Learned

The initial IPDS system was made available on a platform that was far from common among personal computer users (IBM OS/2). This had some known advantages in that even though most PC users could not use the system, we would still be able to get fairly broad usage reports. Electronic mail forums were created and tracked for the user community (approximately 10,000 copies of IPDS were sold in the first year of release). For the most part, this confirmed our belief that at least some audience was eagerly awaiting general-purpose speech recognition technology. A small, vocal community "loved" the product and was willing to work around any deficiencies. However, it also dimmed any hopes that we might have had that the "problems" we had tried to address in error correction and smooth mode transitions between dictation and command had been solved. Had our goal been to develop a niche product, we would have been able to declare success.

While we did not attempt to collect extensive user data for the product, we did collect some detailed use data for a few users that we believe were typical. One of the authors (Karat) worked with the development team on the design of the system. He used IPDS to compose his electronic mail for a period of about six months in late 1993 through early 1994. During this time he collected personal statistics on performance for this task and compared this to his own performance for electronic mail composition using a keyboard. Briefly, the author was slightly faster in creating correspondence with IPDS compared to keyboard entry for the complete task (from starting a message, through editing it, to sending it) - about 10-15% faster overall. However, at the end of the six-month period, the author returned to keyboard use for e-mail. The technology seemed fine for the creation of short pieces of informal text, but was not satisfactory for longer pieces (such as multi-page reports or other documents that might be edited extensively). In short, speech recognition technology seemed at the point of providing some benefits for some tasks over keyboard entry, but the overall level of integration still seemed to be lacking.

3. STORYWRITER

About the same time the development of IPDS began, a different group began exploring the development of a more narrowly targeted dictation application. Our target audience for this application was newspaper reporters who suffer from repetitive stress injuries (RSI). This project began because one of our development partners, the technical management at a large metropolitan newspaper, noted an increased incidence of injuries to nerves of the hands, arms and shoulders among the reporters on the newspaper's staff. Such injuries form the symptom constellation for a type of RSI that is observed in people who spend prolonged periods of time using keyboards.   The technical management office looked to the then emerging large vocabulary speech recognition technology as a way to provide their writers with a completely hands-free writing tool.

However, our initial observations of the reporters' use of an operational prototype document creation application that we adapted from the then recently released IBM Speech System Series (ISSS) product, led us to conclude that we could not provide the newspaper with a speech-only editor that would be efficient to use. The root of the challenge was the type of writing these reporters do. They write long, complex stories that require sophisticated editing tools in addition to a means of inputting text. The prototype we first built for the reporters provided them with only a limited function "dictation window" where text input was to be directed. This text area (the basic speech window which served as the basis for the IPDS application) provided the full support necessary to provide the user with speech playback and alternate word lists, both of which were useful for the correction of speech recognition errors. However, it did not provide sufficient support for editing text. The lack of editing tools rendered the application unusable for the reporters' writing task, which was heavily dependent on reworking text.

Both of the applications we built, ISSS for Journalists (our initial prototype) and StoryWriter (the redesigned final version), were based on IBM's first commercially available large vocabulary dictation system. IBM Speech System Series (ISSS), introduced in 1992, was a discrete dictation system that included a fixed twenty thousand-word vocabulary and the capability for a personalized vocabulary of two thousand additional words. For this project, we customized the base product to create ISSS for Journalists by replacing the general purpose "office dictation" language model and vocabulary with ones that reflected the sentence construction and vocabulary usage found in printed newspaper stories. This language model and vocabulary, along with the speech recognition processor, were also used in the StoryWriter editor.

We created StoryWriter to provide our pilot subjects with the necessary editing capability in something other than a keyboard intensive form. StoryWriter (Danis, et al., 1994) is a multi-modal, text creation and editing application. It is not a full function word processor; rather, it is an editor geared to the text creation needs of reporters. We focused on providing the editing functionality the reporters needed in writing, but excluded all but the simplest formatting functions (paragraphing, capitalization) since formatting for publication is an editor's job.

The decision to create our own text editor, rather than to adapt an existing editor to speech commands, was not taken lightly. People often describe their preferences for text creation tools in quasi-religious tones. However, as it was not possible for us to add the necessary speech supports for error correction to existing word processors, we made the decision to develop a new editing application. The resulting editor included the capability to operate on both arbitrary strings of words and linguistic units, for example, sentences and paragraphs. StoryWriter supported the basic operations of selecting, moving, copying and deleting text. It was beyond the scope of the project to support other useful capabilities available in the reporters' keyboard-operated system, such as the ability to display a history of one's own and an editor's changes to a story.

StoryWriter accepts text input by speech or keyboard and commands by any of five input methods. These are keyboard, speech, mouse, foot pedal as replacement for a mouse click, and a new technique we called "point and touch". The technique of "point and touch" used on-screen "buttons" that represented the elemental editing functions (e.g., cut, copy, delete) that could be activated by pausing a mouse pointer over them. We developed a syntax of movements to accomplish some of the standard operations, such as moving text.

We made all modalities for text input and editing available at all times in order to allow users to switch among them as they desired. This design feature allowed the reporters to use the most efficient input means that their RSI symptoms allowed them at any given time.

3.1   Users/Task/Context of Use

User Characteristics. The original target users for our applications were twelve reporters at a major daily newspaper in New York City who suffered from various degrees of RSI. Half of the group continued to write news stories despite their injuries, but the remainder had been reassigned to non-reporterial jobs. The alternate duty assignments continued to involve complex writing, but removed some of the time pressures that exist when one is writing under daily production deadlines.

RSI is a label used to refer to a wide range of impairments that result from an inflammation of the nerves in the hands, arms and shoulders which arise from repetitive movements of the hands (Hagberg and Rempel, 1997). Stress is thought to be a contributing factor in the development of RSI. It affects all aspects of the injured person's life which involve hand movement, including grasping objects, opening doors, combing one's hair and using a standard GUI for text production. With respect to computer usage, some of the reporters in our study were able to use a mouse for pointing and clicking but unable to use a keyboard, while others were not even able to move the mouse for use as a pointing device. The course that RSI takes in an individual varies both within a day and across days. Many of our users experienced a gradual increase in pain and a corresponding decrease in ability to use a GUI text editor as the day wore on, but their disorder did not have a predictable course over time. Rather, they experienced "good" and "bad" days with no apparent pattern.

Task. The reporters we observed used a style of writing they called "tinkering". In "tinkering", editing does not form a distinct stage that **follows** text input as it does in the style called "drafting" that is frequently observed in the composition of simple, short text such as e-mail and letters. Rather, in "tinkering," editing changes are made by the writer at any point in the output process. A typical writing episode begins by the reporter typing a few words, then backing up the cursor to correct a spelling mistake or to replace some of the original words and then inserting new words before completing the sentence. The process is repeated for subsequent sentences but now, in addition, sentence order can be changed and transitions need to be rewritten to make the sentences form a coherent paragraph and story. Watching the process soon dispels any notion one might have that publishable quality prose springs complete from the fingers of the professional writer. Instead, the writers we observed need to craft the expression of their thoughts to produce the smooth, coherent, and captivating text that is considered publishable by their editors.

Context Of Use. The newsroom in which our user population works is very noisy and crowded. A typical workspace configuration is four desks, all facing inward, separated by four-foot walls. The newsroom is densely packed with these clusters. The noise level is quite high, composed of conversations among workers, televisions that are used to monitor the day's events, one-sided conversations as reporters conduct interviews on the telephone, and frequent ringing of telephones. It is a testament to the writers' powers of concentration that they manage to produce high quality prose under daily or weekly deadline conditions in such an environment. The reporters we worked with described these as stressful but not atypical newsroom conditions and noted that they were used to them.

3.2   Capabilities and Constraints

The characteristics of speech and speech recognition technology that presented the most serious constraints in our application design and development effort were: 1) the presence of recognition errors; 2) the time lag that separated the spoken word and its appearance on the screen (which was described above); and, 3) the unsuitability of speech alone for localization of objects. These three characteristics created an interrelated set of design challenges we had to address in our design of the StoryWriter application.

3.2.1   Recognition Errors

The possibility of errors is an inherent characteristic of recognition technology. The use of close proximity, noise canceling microphones seemed to eliminate any significant contribution of ambient noise to error occurrence. The noise level was fairly constant from moment to moment, and therefore predictable, lending itself to algorithmic intervention. Users "trained" the speaker-dependent system in their work environment in order to match their work conditions. This involved reading a specially formulated 1500-word script that provided the speech recognition system sufficient samples of all of the sounds in the English language. A special processor analyzed this sample and generated a description of the user's voice which was used for purposes of recognition. Since the noise level built over the course of the day until daily publication deadlines passed at 6:30 p.m., many of the reporters split their system training session into a morning and an afternoon session so as to make their recording under representative noise conditions.   Nevertheless, some of the reporters experienced disappointing recognition accuracy rates. These varied from the low 80s to the mid-90s.

Recognition errors in text are certainly an annoyance to users of a speech-based text tool, but errors in commands can be very destructive to people's text. Such errors can be especially damaging in cases where people do not watch the screen while dictating and therefore do not catch the mistakes until much after they occur.

An early study with insurance inspectors who were trying to ease into dictating by reading from their handwritten texts (they had previously depended on professional transcription of their handwritten text), showed that text could be corrupted almost beyond recognition by mistakes where text words were misrecognized as commands.

3.2.2   The Problem of Selecting an Object by Speech Alone

It is very difficult to select an object from among a set of like objects by speech alone.   In face-to-face conversation we often use demonstrative pronouns (this, these, that, those) in combination with pointing for efficiency: "give me that (+ pointing gesture) round doughnut". Without the use of pointing, one useful strategy is to bring our conversation partner's focus to the general neighborhood where we are focused ("give me a round doughnut") and then refine our specification ("the one to the right of the smashed one", "sorry, *my* right, your left").

To correct a word entirely by speech in the first application, users had to specify word location relative to their current location: "up five (lines), right four (words)". This was not a satisfactory solution as it was either difficult or slow in execution. It is difficult to accurately count lines and words on a monitor, especially if one does it without the use of the finger for pointing, as was typical for our pilot users with RSI. The alternative strategy is very slow because it requires users to issue several commands as they approximate and then refine the location of their target ("up five, up three, right nine, left two").

3.3   Design Considerations

The three constraints just discussed combine to make writing that includes significant amounts of editing difficult to do efficiently by speech.   In our observations of reporters writing on a word processor with a graphical user interface it was clear that two critical components for an efficient writing interface were quick and easy access to any location in the already typed text. By way of contrast, writing by dictation into a tape recorder is considered difficult for complex text because the lack of a visual representation effectively limits access to what can be remembered and referred to by the writer. A speech-enabled editor was preferred by reporters who had previously dictated into a tape recorder because it made significantly fewer demands on memory. The remaining significant problem is that the *process* of access is mediated by an error prone tool. The degree of uncertainty that speech recognition inserts into the writing and editing process was one of our main foci in the design of StoryWriter.

StoryWriter instantiates a number of design solutions to this problem because of the varied needs of our target population. Therefore, we present here the "best of breed" we were able to design for users at various degrees of impairment, ranging from a complete speech solution to one in which other modalities are used in order to augment speech where it does not work satisfactorily.

All-Speech Solution. In working with various user groups, we have repeatedly found that misrecognition of commands is more costly to users than misrecognition of text words. A particularly clear example of this is when some word is misrecognized as a vocabulary switching word and all subsequent words, until the error is detected by the speaker, are decoded against the incorrect vocabulary. An example of a problem that plagued the users of StoryWriter in the early stages of development was when a text word was mistakenly decoded as a cursor repositioning command. The resulting text would be very confusing to our users because sentences would have segments of other sentences inserted within them. Even though we had implemented a powerful "undo" mechanism, our users found the problem so baffling that they typically discarded the text from the point of the incorrectly embedded sentence.

Moded Interface. One solution adopted by some of our users was to work in a mode where the command vocabulary was separated from the text vocabulary. Such moded interfaces have the disadvantage of adding time consuming overhead for writers who "tinker" and may want to change a single word before continuing to enter text because two vocabulary switching commands are needed in addition to the commands needed to make the change. This type of solution is favored by some users because segregation of the types of words does make errors less likely. In addition, working in this moded environment has the side effect that the user focuses on the screen while making editing changes and therefore is in a position to detect errors as they happen. Therefore, errors tend not to cascade as they do in unmoded interfaces where text and command vocabularies are combined.

Unmoded Interface. A single vocabulary interface was more suited to our users who, as we have said, preferred to write by "tinkering." For this group we worked on decreasing the likelihood of *false-alarm* errors, where the command is decoded even though it has not been spoken. (The case where the command is spoken but *missed* by the recognizer is less damaging to the existing document.) We created multi-word commands that had acoustic patterns unlike text words and therefore would be less likely to be decoded when the speaker had not uttered them. Commands such as "cursor-up-five" were successful in decreasing the likelihood of false alarms on commands, although they were still confusable with other commands. For example, since we had to implement a set of cursor movement commands that only differed by the last syllable, there were often within-set misrecognitions.

One caveat for this type of solution is that it can work reliably only with applications based on discrete speech recognition. In MedSpeak (to be discussed later), which uses continuous speech, hyphenated multi-word commands are not as easily distinguishable acoustically from text. One may be able to create commands that are highly unlikely to appear as sequences of text words, but this forces the language model alone to detect the anomalies, and loses the power of the acoustic processor.

Multi-Modal Interface. Command execution by one of the non-speech modes, whether the keyboard, the mouse, the foot pedal, or the "point and touch" mechanism, gave users a high degree of correct completion. In a "Wizard of Oz" study that compared four methods of executing commands, users preferred using speech commands for basic editing functions, but would accept use of a mouse click or key press signal if nearly complete accuracy by speech could not be assured. The speed and efficiency of the non-speech methods was very dependent on the method and the user. Keyboard editing was fastest and most efficient for those who were able to use it because they were most practiced in it. One of the patterns we observed was that users were selective in which input modes they used under which circumstances.

The Problem Of Selection Of Objects. We noted earlier that selection by speech was cumbersome in our system. We tried and eliminated some alternative speech implementations. For example, we tried numbering text lines, which would have allowed more direct access to an editing target (e.g., "line-25" instead of perhaps "cursor-up-eight", "cursor-down-two"), but it took up screen space that the reporters felt they needed for display of their text. It also required creation of another set of commands that shared similar acoustic properties, which created a difficult recognition task.

However, selection could be accomplished easily and accurately by moving a mouse pointer over a word and either clicking the mouse, or depressing a foot pedal, or by speaking a command (e.g., "select-word," "delete-word"). The combination of pointing and speaking solved a number of problems.

Lag. In addition to decreasing errors and increasing the efficiency of the selection operation, selection by pointing and speaking circumvented the problem of lag for command execution. This was especially noticeable when reporters worked under unmoded conditions. Recall that the design of the speech recognition technology imposed an additional source of lag because the language model component analyzed the two words subsequent to a given target word in order to make a recognition decision. Trigrams are not used in command recognition mode, because we do not have a good language model for this mode of operation. However, when working with a combined text and command vocabulary, the recognizer would pause in a tentative recognition state if the user paused in order to ascertain correct recognition before continuing dictating. The only way to force the recognizer to make a recognition decision on a command without uttering additional words was to turn off the microphone and turn it back on.

The problem of lag in text creation could not be addressed by design decisions for the interface. In most cases our users preferred to look away from the screen because they found the lag distracting. One problem remained if users needed to look back on dictated text for guidance on how to formulate the continuation of their sentence if it had not yet been displayed on the screen. In these cases, turning the microphone on and off forced recognition to conclude and the text to be displayed.

3.4   *Lessons Learned*

The StoryWriter design and development team had to address two major issues. First, StoryWriter had to be usable by a population of users whose physical impairment due to RSI precluded some of them from using their hands to manipulate either the keyboard or the mouse at all. Second, the StoryWriter application had to be useful for a type of writing we called "tinkering," which is defined by frequent switching between dictation and editing functions.

Our success for the first design challenge, a completely hands-free interface, was limited. StoryWriter was used in that mode, but recognition errors made the experience frustrating for the users. Given limitations in recognition accuracy, self-reports indicated that using the application in its moded version, with text and command vocabularies activated by vocabulary-switching commands, produced the highest level of success and satisfaction. One would expect that improvements in the recognition technology that increase the accuracy rate will make a hands-free speech interface more usable in the future. However, as the technology improves and continuous speech becomes the norm in recognition technology, an all-speech interface will have to continue to be moded in order to be able to distinguish reliably between text words and commands.

StoryWriter was much more successful when speech text input could be augmented by non-speech command entry. StoryWriter implemented a number of command entry procedures, some of which required minimal hand movements. The feedback we received indicated that this made the application quite usable for most of the reporters with RSI. Since one did not have to indicate, other than by usage, one's choice of a command entry method, users were able to flexibly move among the methods as needed.

Addressing the second major challenge, that of supporting the editing-intensive writing task of reporters, revealed some important design considerations. Once again, an all-speech interface was not found to be the most usable one according to self-reports. Part of this problem will also be addessed by improved recognition performance that will make command recognition more accuracte. However the crux of editing is the selection of words upon which editing operations are performed. While one can conceive of an all-speech interface for word selection, the process will not be as simple and easy for the user as simply pointing to the word which is the editing target.

Thus, the major lesson of the StoryWriter design experience is that we must pay attention to the "affordances" of the technology we use in applications. While an all-speech interface is sometimes necessary (e.g., for users with RSI, over the telephone, in hands-busy situations) and is possible, augmentation by other modalities, especially for editing functions, results in a much less distracting and useful interface for users.

4. MEDSPEAK

This section describes a speech recognition application that allows for the transcription of dictated speech into text. Unlike the applications described in the two preceding sections, this product was created for a **specialized group** of users (radiologists) and uses **continuous** recognition technology (which allows users to speak without a discernible pause between each word). Beyond these two major differences, MedSpeak shares much with the other two IBM applications described thus far. They were all comprised of a system with a monitor that displays the GUI, a mouse and keyboard, as well as a microphone and set of speakers. The user identifies himself or herself to the system which loads that individual's personalized speech profile (macros, audio settings, and enrollment if there is one) as well as other user settings such as font size and function preferences.

MedSpeak/Radiology is a commercially available speech recognition application developed in 1996 that supports dictation, editing and electronic signature of radiology reports (Lai and Vergo, 1997). It uses the IBM Continuous Speech recognition engine as of late 1995 to early 1996, which was both continuous and speaker-independent. The predominant form of input is by voice for both dictation (verbatim transcription) and navigation within the application (e.g. "View Preliminary Reports", "Exit Application"). Radiologists can create and save frequently used text, recalling it later with the utterance of a single word or phrase. Native English speakers that we sampled achieved an average accuracy percentage in the mid-nineties. People with heavy accents or who do not speak clearly needed to train MedSpeak to the sound of their voices in order to achieve or approximate equivalent accuracy rates.

4.1 Users/Task/Context of Use

User Characteristics. The user group for our product consists exclusively of radiologists. Radiologists are highly specialized doctors, responsible for interpreting X-rays, MRIs and CAT scans and compiling the information into a patient report that goes back to the referring doctor. We worked closely with 22 radiologists at two large medical institutions and one small community hospital. We conducted studies with a total of 40 radiologists, including the 22 previously mentioned. Many of the doctors with whom we worked spoke English with a pronounced foreign accent, which became a slight challenge in terms of having them achieve high accuracy with speech recognition.

Tasks. Most radiologists are paid per report. Due in part to that fact, the process for creating reports today is very streamlined to minimize the radiologist's time since this is the most "expensive" component in the process. Prior to MedSpeak, the process for creating a report almost always used a tape or digital recording system. The radiologist would pick up a microphone, state who he or she is, who the patient is, and what the procedure was. The radiologist would then dictate his or her interpretation of the film/s. The doctor can remain totally focused on the film since the mechanics of the tool for capturing the dictation are very simple.

Context Of Use. Radiologists dictate in "reading rooms". There are usually at least two radiologists dictating in a reading room, and there can be more depending on the size of the hospitaland the time of day. The rooms are large spaces with several light boxes with alternators in use. In addition to the sounds of the radiologists dictating, telephones ring, alternators make noise as they switch the set of films currently in view, side conversations take place, and doctors are paged. While this introduces a fair amount of noise, it is inconsistent in duration and intensity. Early in the morning and late in the evening one can often find a sole radiologist working alone quietly and efficiently.

4.2   *Capabilities and Constraints*

The speech engine used in the initial version had a speaker-independent model for speech that had been built with only 40 prototypes, most of them from around the New York area. A prototype can be thought of as a sample of speech, and the SI model is built from the merging and "averaging" of these samples. Thus, if one spoke like a typical New Yorker and was able to produce sentences with a fluency comparable to that of read speech, one could achieve accuracy similar to what was measured in our labs. This was not the case for many of our users.

The next iteration of the software had a model based on 600 prototypes. As a result, the speech engine could recognize people speaking with a southern twang as easily as someone from Seattle or New England. This resulted in a dramatic increase in the accuracy of the engine between the start of the project and its finish. One study showed an average error rate of 7.62% during an early stage, which dropped down to 2.97% for those same users by the time we had reached a finished product.

Another problem that we encountered in the early phases of the project was the difference in accuracy rates between read speech (handing a radiologist a printed report and asking him or her to read it to the system) and spontaneous speech (when the radiologist is composing on the fly). The latter speech results in many "umms" and "aaaahs", as well as both elongation and shortening of syllables. Additionally there are usually false starts and random insertions of text due to breath noises and lip smacks. All of these increased the error rate and the user's frustration with the system.

Initial versions of the system were not robust to silence at all. If the microphone was turned on in Dictation mode, any small sound including the scrape of a chair on the floor would be turned into a word. This proved to be problematic for our users since they would often pause during their dictation to examine the film more closely. While in Command and Control mode we employ a "mumble guard" that ignores sounds it does not recognize to be active commands, and this feature works well.

4.3   *General Design Issues*

Our task as designers was simplified by the fact that we were creating a tool that was to be used by a narrowly defined set of users whose characteristics were fairly well known to us. MedSpeak needed to be a tool that users would use all day, every day, and we did not have to achieve "walk up and use" ease of use. The usability goal for our product was to create a system that could be used with two hours or less of training with report creation times comparable to those of the current system. These goals were driven by the fact that many physicians exhibit a reluctance to use computers and thus would be unmotivated to spend much time training. Also, radiologists are compelled to produce reports rapidly. A system that required significantly more of the radiologist's time to create a report would not gain much acceptance because the radiologists would feel the financial impact of the additional time. We focused our efforts on providing a simple, fast path through the application to create a radiology report.

We needed to make a system that would not be intimidating to first-time users, but would stay out of the way once the user had found his or her way through system. All new users received a default profile that caused the user interface to display only the basic function set. Basic functions included everything necessary to dictate, edit and sign a report. Advanced functions included things like exporting the report in ASCII, creating voice macros, and adding a personal pronunciation for a particular word. Functions were not organized into a menu structure but were accessible through large push buttons because we wanted all primary functions visible at all times. We opted to have the user decide when he or she was ready to "graduate" to a more advanced level of functionality. Through exploration, the user could discover the setting for the advanced functions in his or her customization panel. If advanced functions were enabled the user would have a second row of push button-accessible functions available on the interface.

The push buttons could be activated by voice, or by mouse as a fall back position. Their existence served as a reminder to the user of the functions that could invoked by voice. Because we knew there would be times when the user would have a hard time getting the system to recognize a voice command (e.g., too much background noise, wrong microphone position, or the user had a cold), we needed to have either a keyboard or a mouse means of activating every command.

We opted for push button (tool bar) access to functions with no menu bar because the push buttons are a large easy target for naive mouse users and because they provide the added value of being a constant reminder of what functions are available. The problem we ran into was that while the user saw the button and remembered the functions, new users often forgot the voice command for invoking the button. This problem was further aggravated by the fact that we needed long command names in order to maximize the probability that the voice command would be correctly recognized. For example, the command to dictate a new report was "Dictate New Report," and the one to view a listing of the reports previously dictated by the doctor but which had not yet been signed was "View Preliminary Reports." These commands were long enough to be unambiguous for the recognition engine, but were too long to display the text of the command on the push button. Thus, the user either needed to remember the particular command or issue the "What Can I Say" command that would display all the spoken commands active at the time.

One freedom we did have in designing MedSpeak is that we did not need to follow known word processing standards because there was no common knowledge upon which to build in our user community. Even though we were basically building an editor, very few radiologists were familiar with either Word or Word Perfect so we were not tied into incorporating a menu bar with the standard File Edit View configuration. Interestingly enough, while we did not encounter any desire from the radiologists to conform to known Windows '95 practices in GUI design, we did need to overcome some resistance amongst the programmers on the team who were uncomfortable with any such deviations. "What no Menu Bar?!"

In order to gain the highest initial acceptance level possible, we modeled the application flow to reflect the current work flow process. Thus, doctors who were used to dictating a series of reports before reviewing any of them could continue to do so, even though the preferred model was to dictate, review and sign. This model was preferred from an overall throughput viewpoint because this got the reports out into general circulation as fast as possible. It also gave the radiologist a sense of closure that increased satisfaction for many radiologists.

4.4   Speech Recognition Design Issues

Recognition Failures. The state of the technology that we were using was such that the engine never returned a recognition failure when the system was in dictation mode. It always made its best guess. There were times that it clearly should not have tried to guess, but this was not a parameter over which we had any control. However, when in command and control mode, it was possible for the engine to fail. In other words the acoustic signal being received by the engine could be so different from any command that the system had stored, that we returned a "Try Again" to the user. Since it was possible for the engine to fail repeatedly in trying to recognize a command, we used Try Again with a counter (i.e., Try Again [1], Try Again [2], etc...) so the user at least could tell that something had been heard. While we felt this was a less than optimal solution, the time constraints that we were operating under did not allow us to implement a more demanding change to the interface such as tailored and incremental messages to the user. The two most recently recognized commands displayed in what we called the Command History area. This was a small window on the toolbar in the upper left-hand corner of the screen. While we believe it was sufficient to show only the last two recognized commands, it would have been helpful if this area had had more real estate because we were constrained in what we could display in the window by its width.

Recognition Errors. Something that happened less frequently than a recognition failure in command mode was a recognition error (e.g., the user says "Dictate New Report" and the system understands "View Preliminary Reports"). This is partly due to the fact that the active vocabulary for the command set is fairly small, and also because we chose long command names to optimize our chances of correct recognition. If the wrong command was recognized the user could undo whatever had been done with the undo command. If the command was destructive (e.g., "Start Over," which deleted the entire report) or permanent (e.g., "Accept and Sign," which electronically signed the report and distributed it), we would have the user confirm his or her request before taking action.

Latency. Our design decision was to not show infirm words. In an earlier version of the software which displayed infirm words and then changed them as the engine firmed them up, we found that radiologists glancing over at the screen were distracted when a mistake showed up, even if this mistake was later adjusted by the recognition engine. This distraction often caused them to lose their train of thought. Latency was not much of problem for us. Our primary requirement was to have the transcription complete by the time the doctor said "End Dictation" and turned his attention away from the films and back to the computer screen. Of course there was variation among doctors and between reports in how easily dictation was accomplished. However, it was clear that the dictation task was easier for radiologists than it would be for people who had not spent considerable time dictating into a tape recorder.

Error Correction. If the user spotted a recognition error, he or she needed to invoke our error correction dialog to correct it. This dialog was designed with the optimistic viewpoint that the alternate word list (words that the recognition engine had ranked as being likely alternatives to the word that it chose as its first choice) would likely contain the word that the user had actually said. However, the technology at the time was such that the alternate word list almost never contained the word the user wanted. Thus, correction time was significantly increased because the user scrolled through the list looking for the word, then usually had to type it in anyway. The reason we did not have the user just say the word again is that it was most likely the engine would get it wrong again. This was not always true, but did occur more frequently when the user was fairly new to the system. The error could have been due to the word being out-of-vocabulary, in which case the engine did not have a chance of getting it right no matter how many times the user said it. Alternatively, the error could have been due to the fact that the user pronounced the word differently from the pronunciation that the system had stored for the word. In these cases, the user needed to add the word and record a pronunciation for it or, if it was a word that already existed in the vocabulary, to store his or her pronunciation for that word. All these functions were available in the correction dialog.

Our primary problem with the early correction model was that this dialog was only necessary when the speech engine made an error. It allowed the engine to "learn" how the user pronounced words and improve its performance for that user. However, when a user wanted to make a change to the text simply because he had changed his mind about how he wanted to say something, he only needed to highlight the text and redictate it, or to type in the new text. No use of the correction dialog was required. Our users had a hard time understanding this difference. This difficulty was compounded by the fact that there are different types of recognition errors. When a substitution error occurs, the engine simply replaces one word with another. But sometimes it replaces one word with two or three words, and all of these words need to be selected for correction through the error correction dialog. There are also insertion and deletion errors. With an insertion error, the engine includes a word when the user did not say one. This could be due to breath noise, or a slight disfluency on the part of the user. A deletion error is when the engine drops a word. Because we found these nuances too subtle for our users, and our users were confused as to when to use the correction dialog, we ended up with a design that did not use a correction dialog at all. We did away with the alternate word list since it was not doing us any good. When the radiologist electronically signed his or her report as final, the program quickly scanned the report for new words and prompted the user to record a pronunciation if it found any words that needed to be added to the vocabulary. The ability to record a new pronunciation for a word that was often misrecognized for that user was relegated to the advanced function set.

Feedback of State. It was important for the user to know rapidly and easily what mode he or she was in: dictation, command or paused dictation. Since the light box for viewing the films was usually not directly adjacent to MedSpeak, the radiologist needed to know the system state from a glance across the room, or out of the corner of his or her eye. In order to do this we relied heavily on strong visual cues through the use of color. The major portion of the screen for report creation consisted of the area that accepted and displayed text during dictation. The background color of this area (technically the multi-line edit widget) was gray in command mode, changed to blue during dictation and bright yellow when dictation was paused. We also grayed out all the push buttons during dictation except the "end dictation" icon as none of the other commands was active. Since there was a delay of a few seconds between the time the doctor said "begin dictation" and the time the engine was actually ready to accept dictation, we added an auditory beep so that doctors would know when dictation mode had been initiated without actually taking their eyes off of the film. This combination worked well during the field trials. The default state for the application was command mode; the user moved in and out of dictation mode with a verbal command (or button click).

Eyes-Busy/Hands-Busy. One of the goals we had for the design of the system, based on what we had learned from the workflow analysis, was to ensure eyes-free operation during dictation. Radiologists were concerned that if they turned their attention away from the film they might miss something in it. This goal was supported by the technology and aided by some auditory cues that we added. The users also would have preferred hands-free operation, for which we had initially strived, but fell short of delivering. Certain areas of the workflow could not be navigated with voice alone based on the technology we had at the time. For example, digit recognition was poor enough that we decided to require that the patient medical record number be typed in. Also, navigation within a dialog was awkward so, if the user wanted to change the settings on his customization dialogue, for example, he or she had to do it with mouse and keyboard.

Enrollment. Many of our users enrolled during the early phases of the project to improve their accuracy. Some of these doctors were non-native speakers who spoke English with a pronounced accent. In all cases except one, enrollment allowed the doctors to achieve average accuracy rates of 90% or above. One radiologist who spoke with a thick Indian accent was only able to get around 85% accuracy even with a long enrollment. With a 15% error rate she found that using the system to do her daily work was somewhat frustrating and much slower than her current method of transcription.

4.5   Early Use of the System

We had three field trials with different iterations of the software before the product was announced. During these trials, important usability issues were uncovered which we attempted to fix before going out with the next iteration. During the first field trial we found that the recognition accuracy was less than we had hoped. This was due primarily to speech issues we had with the engine that were later addressed.

We learned, however, that below a threshold level of accuracy (which was around 90%) the doctors would not use the system at all. This prevented us from getting any feedback on the usability of the application and caused us to concentrate our efforts on improving the recognition accuracy of the system.

Another issue that became apparent to us during the field trials was the importance of full integration with existing systems. Successful completion of the task was not sufficient for our users unless it was interfaced with legacy systems. During the first two trials, even though the radiologists were able to dictate a report and print hard copy for manual distribution, they were not willing to use our system for "real" work. They still felt that it was relegated to the toy category, and would take time out of their busy schedules to use the system only when we were on-site with them. They were accustomed to having the report available in soft copy on their Radiology Information System (RIS) and would not use MedSpeak to do their daily work until it was fully integrated with the RIS.

We opted against having "inline" commands. These are commands that are active during dictation (without switching into command mode) so if the user speaks them, the engine will recognize that it does not need to transcribe the words verbatim but needs instead to act upon them. By the end of the trial phases, the only command we had active during dictation was the "End Dictation" command that returned the user to command and control mode. We experimented with having "Go to Sleep" as an inline command. When spoken, this command puts the microphone to sleep until it hears a "Wake Up" command. A pause function was necessary since radiologists often pause during dictation to examine the film more closely and, as previously mentioned, early versions of the software were not at all robust to silence in dictation mode. The users' preference was to be able to pause with a voice command so that they could keep their full concentration on the patient's film. The problem we encountered was that "Go to Sleep" was frequently misrecognized when the user was dictating text. This problem was compounded by the fact that the user had no auditory feedback that the system was asleep, and thus would continue dictating without realizing that none of the text was being transcribed. We briefly tried auditory feedback in the form of a recorded message that stated "The system is asleep". While users were now informed of the change of state, they were no more pleased with it. We eventually took out the inline command for sleep, and incorporated a pause function that the user could toggle on and off with the space bar.

The pause function was also helpful in dealing with one of the other issues that surfaced during the field trials. Users were able to achieve the highest level of accuracy if they had the text they wanted to dictate well formulated in their heads before speaking. This reduced the number of false starts and other disfluencies attributable to spontaneous speech. With an easily accessible pause function, we found our users composing a thought or two, speaking and then pausing the system again to compose the next thought.

While we had concentrated on making sure the user had correct feedback as to what mode the system was in, we found during the trials that there was insufficient feedback to the user when the command was not recognized by the system. The only indication of such was that instead of acting on the command and displaying it correctly in the Command History area, the system displayed "Try Again" in the

Command History. There was even less feedback with repeated failures since the only indication was that the counter for "Try Again" was incremented. This was very easy to miss at a distance. Until the users learned to watch the Comand History to see what had been recognized, they were often left waiting for something to happen. Stronger feedback of a recognition failure would have been helpful. Since we had relied on the use of color for feedback of state, we think now that we should have tried changing the background color in the Command History when recognition failed.

We learned rapidly that many members of our user community were demanding and, at times, impatient. Like many busy people, they had little tolerance for an imperfect system. When we would take a new version of the software to the client sites (new software installations were usually separated by an eight to ten week time span), the doctor would try the system and, after blasting through some dictation, would turn to us and say "see, it still makes mistakes.". Several of the doctors in each group, however, could be viewed as "early adopters," and they would be more prone to view the same example and say "wow, it only made two mistakes". These early adopters were our strongest allies. They won over other doctors with their enthusiasm and gave freely and generously of their time, providing us with valuable feedback and suggestions for improvement. However, we have sometimes wondered if these early adopters should have been excluded from the final evaluation of the system because, in many ways, they were no longer objective. After months of working with us and seeing their suggestions incorporated into subsequent versions, MedSpeak had become their baby too, and they were probably more tolerant of its shortcomings.

Since the amount of time spent on a report often has a direct correlation to the amount of money a radiologist earns, this user group always seems to be functioning in fast-forward mode. The expectation when they used the system for the very first time was that they could speak the same way they speak into their dictation machines. However, we found that there was fairly rapid two-way adaptation that went on between the person and the machine. The machine would update its statistics on the user every time the doctor signed a report, thus improving its performance for that user over time. In many cases, the doctor appeared to adapt faster than the machine. During field trials, we found that the first time doctors dictated to the system they would dictate in the same manner they had always dictated - speaking very rapidly and softly, often running words together. They would watch the screen as it displayed the recognized text, including the errors resulting from the manner in which they had spoken. Then, like children with a new toy, they asked eagerly if they could try again. Within the first three or four attempts, their accuracy usually jumped significantly as they adapted their manner of speaking to a manner that gave them a higher "success" rate.

One of the biggest causes for low accuracy during the trials was poor microphone usage. Either the volume was too high (user spoke too loud or had the microphone too close to his or her mouth) or too low (user spoke too softly or had the microphone too far away from the mouth). We had a volume indicator in the interface, which fluctuated from yellow (too soft) to green (good volume reading) to red (too loud). However, we found that users usually ignored the volume indicator since their eyes were not on the screen anyway. Since we were using a noise-canceling microphone to deal with the environmental noise in a reading room, microphone position was sometimes a problem as well. Unless the user spoke down the long axis of the microphone, the recognition was degraded since the microphone tends to cancel out sounds coming in from the sides of the microphone. Radiologists who were used to speaking into a Dictaphone and holding it straight up and down had to make an adjustment to hold the MedSpeak microphone positioned diagonally from their mouth.

4.6   Lessons Learned

In general, we were very successful. Radiologists were able to complete their reports in a much faster cycle than with traditional methods (Lai and Vergo, 1997). We did not have the time to implement all the solutions we had in mind. Table 2 shows a listing of both the solutions we adopted, given the time and technology that we had, as well as what we would have chosen to implement given more time. We are working on several of the proposed solutions at this time.

Table 2.   MedSpeak design solutions.

Problem Discovered	Resolution Adopted	Proposed Solution
user spoke too loudly or too softly	volume indicator	intelligent agent that monitors volume and intervenes helpfully
microphone in wrong position	Training	agent as in above - if recognition degrades suddenly one could prompt the user to check the microphone
engine not robust to silence during dictation	pause function controlled with spacebar	"speak to talk" button on microphone
new users forgot voice commands	"What Can I Say" function	Natural Language Understanding
recognition failures in Command & Control	"Try Again [1]"	tailored and incremental messages to the user with strong visual indication of failure

User-centered design was a cornerstone of our development approach with MedSpeak. It was critical for us to get a "usable" version of the software out into the hands of radiologists so that we could observe them at work with the prototype and conduct interviews with the doctors. Since MedSpeak was required to be a system that radiologists would use for long periods of time during the day, every day, we felt certain that some of the usability issues would become apparent only with protracted use. However, we quickly found during the first and second trials that the doctors would use the system only when we were on-site and had scheduled time with them. The rest of the time they fell back to their normal routine which did not involve the use of MedSpeak. We believe there are several reasons for this reticence to change routines, some of which pertain solely to the introduction of new technology in an existing workflow, and others which have a direct correlation to the introduction of a speech system.

Below a threshold level of accuracy, which was as low as 90% for some doctors but closer to 94% for others, radiologists were not interested in investing the time to create a report with speech. In some cases, the bumps in the road were due mostly to the user being new to speech recognition, and with some motivation to continue, usage would have become smoother and painless. Unfortunately, certain doctors were disappointed the first time they tried the system and proved difficult to motivate for a more sustained effort. Therefore, while we wanted to get a system out to users as early as possible, we needed to be careful that we met a minimum level of accuracy with the system so as not to turn away potential enthusiasts.

Our users were very experienced with text creation by means of dictation. Even so, several doctors saw a sharp increase in the recognition error rate due to the lack of fluency in their speech when composing text. While human medical transcriptionnists are able to easily filter out the disfluencies in the doctor's speech, MedSpeak was not. The doctors that adapted most easily to the speech recognition system were the ones who could dictate almost as cleanly and smoothly as if they were reading text from a written report. Application features that support this difficult task of oral composition go a long way towards increasing early acceptance and continuing use of the system. Several of these features are: an easy pause function to compose one's thoughts, a "scratch that" type function, oral selection and correction of text, as well as seamless movement between navigation mode for correction (e.g., "go to the bottom of the document," "playback the third sentence") and verbatim dictation. It is to support this last feature that so much research and focus is going into Natural Language Understanding and Modeless operation.

5.   THE CONVERSATION MACHINE

This section describes a speech recognition application that uses conversational speech as input and output. As with the other applications discussed here, avoidance of and recovery from speech recognition errors is a major user interface design challenge. However, the task, users and context of use posed rather different design challenges and required rather different solutions than for the other three speech recognition applications discussed here. We will focus on the design decisions and, in particular, the role of feedback in the form of interactive prompts intended to facilitate error avoidance and recovery. We will base our discussion on our experience in the design of the Conversation Machine (Wolf, Kassler, Zadrozny, and Opyrchal, 1997; Wolf and Zadrozny, 1998; Zadrozny, Wolf, Kambhatla, and Ye, 1998) and the experience of others in the design of all-speech interfaces (Ly and Schmandt, 1994; Marx and Schmandt, 1996; Yankelovich, Levow, and Marx, 1995). While the application domain was banking, the research objective was to develop design principles which can be applied to many domains.

The Conversation Machine for Banking allows a user to get basic banking information such as account balances and perform a number of transactions, such as transferring funds between accounts and paying bills, over the telephone. For example, a user might say, "What's my checking account balance?" and the system responds, "Your checking balance is $250.35." Below is a sample dialogue (from Wolf and Zadrozny (1998)). For most transactions, there are dozens, or even thousands, of different utterances that can be used. Thus, the system is not simply a continuous speech command system in which one or a few utterances are mapped to a particular command. The goal is to make the user's experience more like a conversation between two people.

With conventional technology, a user would accomplish these tasks using the telephone keypad to make selections from a menu presented by an Interactive Voice Response (IVR) system.

User:	*What's my checking balance?*
CM:	Your checking balance is $925.00. What else can I do for you?
User:	*Tell me about my bills.*
CM:	Which bills do you want to know about?
User:	*Mastercard and phone.*
CM:	Your Mastercard balance is $325.50. Your phone balance is $45.00. What else can I do for you?
User:	*Pay them.*

The Conversation Machine for Banking is more than a research prototype. It was developed in cooperation with a banking partner with the goal of offering a conversational interface as an alternative to a keypad and menu interface for telephone banking. As of this writing, a limited field trial is scheduled to take place in 1998.

The Conversation Machine for Banking includes the following components:

♦ IBM's Continuous speech recognition for telephone—to recognize the words in the user's request

♦ Natural language/dialog processing—which uses semantic, syntactic and pragmatic information to determine which action to perform (Zadrozny, 1995, 1996).

♦ Digitized prerecorded speech output—to present information in response to the user's requests

♦ Computer-telephony integration (IBM's DT/6000)--which handles integration and communication between telephony functions (e.g., touch-tone input, speech output) and computer functions.

All speech recognition and natural language/dialog processing is done on a server. To use the Conversation Machine, the customer simply makes a call from a conventional telephone. The call is routed over the public switched telephone network like any other voice call.

5.1   Users/Task/Context of Use

User Characteristics. Since anyone who calls an automated telephone banking service is a potential user of the Conversation Machine for Banking, potential users are a varied group in terms of age, gender, experience with computers, native language and frequency of use. Some customers call the bank to check on their balances several times a day, while others call only once or twice a year. Even frequent users, however, may be expected to use the system for at most a few minutes a day. Unlike MedSpeak or StoryWriter users, users of the Conversation Machine for Banking are likely to have no time or motivation for training the system. Our experience in a laboratory study suggested that some people will not listen to even a brief introductory message explaining how the system works. The system, therefore, had to be speaker-independent and require little instruction to use.

Task. The Conversation Machine for Banking allows users to perform almost all of the banking functions accessible from an automated IVR system using conversational speech. This differs from the dictation tasks for which IPDS, StoryWriter and MedSpeak were designed. In these applications, the primary goal is to produce text from speech input, whereas with the Conversation Machine, the goal is to accomplish a transaction using speech. This difference in task has implications for what constitutes an "error." Whereas for dictation, an error has occurred whenever the resulting text is not a word-for-word match to the spoken input, for a transaction application such as the Conversation Machine for Banking, an error has occurred whenever the spoken input does not result in the desired action. The design of the dialogue-processing component takes advantage of the fact that an utterance need not be recognized exactly word-for-word in order to map it to the correct transaction. For example, the user might say, "I'd like to know my checking balance." If the system recognizes "Please show my checking balance," the Conversation Machine will perform the correct action. This characteristic has beneficial effects on transaction accuracy; the transaction accuracy rate reported in Wolf et al. (1997) was 20% higher than the utterance accuracy rate based on an exact word-for-word match.

However, because the task involves a transaction, it is possible that a recognition error will change the state of the user's bank accounts in an unintended way, with unintended consequences on subsequent transactions. For example, transferring $500 from checking to savings, rather than savings to checking might cause a check to bounce.   People can be particularly unhappy about mistakes involving their money. Consequently, one task requirement is to make it easy to detect and correct errors before the transaction is executed, and to provide fail-safe methods for immediate recovery from transaction errors. Although uncorrected recognition errors can have dire consequences in dictation applications like MedSpeak, such applications may permit correction to be done in a separate editing pass, whereas a banking transaction system requires error correction to be performed in the context of the transactions.

Context Of Use. Users may want to use the Conversation Machine for Banking from all the environments in which they currently use keypad input, including the home, office, and telephones in public areas. These environments can be noisy and prone to interruptions. For example, our observations indicate that consumers at home often have television and children in the background. What's more, users may attempt to use the system from cellular telephones, which often have audio quality that is inadequate even for human-human communication. The acoustic context of use requires that the system be designed to accommodate potential recognition problems due to noise.

An important distinction between the context of use for the Conversation Machine and the other three applications discussed in this chapter is that the context of use for these other applications includes a visual display. The absence of a display means that all feedback, including feedback associated with error avoidance and recovery, must be presented through the auditory channel.

5.2   Capabilities and Constraints

The state of the art of the technology imposes some constraints on the design of the application and user interface. Continuous speech recognition over the telephone can handle a vocabulary as large as 60,000 words. The limits on vocabulary size for acceptably accurate recognition depend on many factors, including the complexity of the grammar, the possibility of the natural language/dialog management component compensating for recognition errors, and the task, user, and context of use. We have found that the Conversation Machine works best when the active vocabulary at any point is limited to several hundred words. This appears to be an acceptable constraint, given the vocabulary people tend to use for banking, and our strategy of designing dialogues with the goal of reducing the probability of a speech recognition error. Overall, however, the Conversation Machine for Banking can process about 50 million different utterances.

A second characteristic of the technologies used for this application that distinguishes the Conversation Machine for Banking from the other applications discussed in this chapter is the use of natural language/dialog processing in combination with speech recognition. Since many different utterances can be used to accomplish a particular goal, error correction using speech does not depend on the recognition of a unique utterance.

5.3   Design Issues

As stated in the introduction, avoidance of and recovery from errors is a major user interface design challenge. However, in the Conversation Machine for Banking, it is not recognition accuracy per se, but the ability to complete the intended transaction that is important. To a user, an error has occurred whenever spoken input does not cause the desired result. Similarly, when the result is correct, the fact that all spoken words were not correctly recognized is irrelevant (and invisible) to the user. We will refer to these as transaction error and transaction accuracy to distinguish these from recognition error and recognition accuracy.

Transaction errors as experienced by the user fall into three categories.

♦ The Conversation Machine is unable to understand the user's request and does not attempt to carry out any transaction. It may respond with a phrase such as, "Sorry, I did not understand what you said. Please rephrase your request or say help."

♦ The Conversation Machine carries out a transaction other than the one intended by the user. For example, the user wanted the checking balance but got the electricity bill balance.

♦ The Conversation Machine begins or continues a transaction other than the one intended by entering a subdialog in which specific information is required. For example, a user wanted to transfer money between accounts but got into a subdialog which asks which bills should be paid. This third type of transaction error is similar to the second type described above, but is distinguished because the cosequences to the user are different.

The appropriate response to a transaction error depends on the error category, the specific result, and the cause of the error. However, a particular transaction error might have a number of different causes, and the cause of the error is frequently not apparent to the user.

Consider the case in which the Conversation Machine carries out a transaction other than the one intended (the second error category above). As shown in Table 3, the transaction error may have a number of different causes, and the appropriate remedy depends on the cause. Consequently, the design of robust techniques for error recovery that do not depend on the source of error is an important issue.

Table 3.   Error causes and recovery methods.

Example	Cause of Error	Error Recovery
sloppy pronunciation of a word	speech misrecognition	repeat or rephrase request
noise from public address system in airport	background noise causes speech misrecognition	repeat or rephrase request in quieter environment
"What's the damage going to be for my Visa bill?"	use of phrase not in speech recognition or natural language grammar	rephrase request--"How much is my Visa bill?"
"How much was my electricity last month?"	system does not know information requested	Change goal

It is worth noting that good error recovery may be at least as important to user satisfaction as error avoidance. Wolf et al. (1997) found that users' success in accomplishing their goals was correlated with measures of satisfaction, whereas transaction accuracy was not. The measure of goal success was based on whether or not users were able to accomplish a goal, regardless of how many transaction attempts it required. In this study perceived transaction accuracy (78%) was much higher than actual transaction accuracy (58%) and highly correlated with actual goal success (goal success rate = 92%, correlation with perceived transaction accuracy =.81).

5.4   Strategies for Error Recovery and Avoidance

In the process of developing the Conversation Machine, we have exposed the system to potential users in formal and informal user tests, and have developed a number of design principles and strategies for dealing with errors. Several of these are described here.

Feedback For State Is Embedded In All Prompts. One of our design guidelines is that each prompt should contain enough feedback for state to allow the user to determine the current state from that prompt alone. We learned this lesson in an early version of the Conversation Machine for Banking. In that version, the first subdialogue in response to a transfer request was "To make sure I understand the amount of money, please enter the amount..." This worked fine when the user had intended to transfer money. However, if the user had actually said something like "I'd like to pay my bills," and through a speech recognition error the system had produced the above response, the user would not be able to detect the error.

Because the burden is on the user to detect a misrecognition, the system can facilitate error detection by providing explicit feedback for state. Speech interfaces that include a display must also provide feedback for state. However, given the nature of vision, it is possible to provide such feedback in a way that it is available when the user wants it but does not intrude upon the user. For example, consider the use of color in MedSpeak to distinguish dictation, command, and paused dictation modes. Given the characteristics of hearing, a challenge for conversational speech interfaces is to design the feedback so that it facilitates error detection, but is not burdensome by being unnecessarily intrusive or lengthy.

Incomplete Transaction Dialogs Acceptable. Dialogues that contain only some of the information required to complete a transaction are followed up with prompts to gather the additional information. For example, if the system understands a request to pay bills, it responds with the question "Which bills should I pay?" This provides several benefits in terms of error avoidance. Simple requests may reduce planning demands on the user and the consequent disfluencies which often result in speech recognition errors (Oviatt, 1995). The follow-up dialogues have a rather small active vocabulary and are more likely to result in correct recognition than a request at the top level of the system. In addition, if some words in the original utterance are missed, it may be possible for the dialog to proceed without the user repeating the entire request.

Defaults And Suggested Actions. If the speaker does not provide all the information necessary to uniquely identify a transaction, the Conversation Machine for Banking may respond with a reasonable default rather than query the user for additional information. For example, if the user asks for "ATM transactions," without specifying checking or savings account, the system responds with recent ATM transactions on the user's checking account. This makes sense, since in most cases users are interested in ATM withdrawals from their checking accounts. Another technique for error avoidance is to try to anticipate a user's next request based on the current state and present that anticipated request as an option. For example, customers often call with the goal of verifying their checking balance. If the balance does not match their records, they frequently ask for their checking transaction history in order to determine the cause of the mismatch.   When a customer's initial request is for a checking balance, our system asks if the user wants his/her checking transaction history, thus reducing speech recognition to a simple yes/no distinction. Defaults and suggested actions contribute to error avoidance by simplifying the recognition problem and reducing the complexity of the dialogues required to accomplish goals. (Marx and Schmandt (1996), among others, also have used suggested actions.) In MedSpeak, macros perform a somewhat analogous function to defaults by reducing the amount that has to be spoken to accomplish the desired goal.

Guidance For The Next Step In Case Of Error. The strategies of providing feedback for state, defaults, suggested actions, and prompting for additional information help avoid error when the conversation is on the right track. We observed in Wolf et al. (1997) that such guidance may not be adequate to help users figure out what to do next in the case of a transaction error. Consider the following example from an early version of the system:

> *User*:   *I want a transfer, of a hundred dollars...*
>
> *CM*:   Which bills should I pay?
>
> *User*:   *I want a transfer, of a hundred dollars...*
>
> *CM*:   Which bill? Phone, insurance, Mastercard or electricity?
>
> *User*:   *No, no bills {pause}. I just want to transfer money from my savings account ....*

In this example, the user does not know how to deal with the error. In the current version of the system, users are reminded to say "Cancel" or use the star (*) key to cancel when appropriate. Such reminders can reduce frustration and help users get back on the right track.

General Advice. As noted above, typically, it is not possible to give guidance based on the specific source of an error. Consequently when there is a recognition failure, the Conversation Machine provides guidance based on generally useful strategies. For example, since long utterances often fail to be recognized, one error prompt suggests the use of "short, simple requests." Our tests with users indicate that this advice is very effective. In addition to helping the user with the immediate problem, this advice may also guide the user to a more successful speaking style. This technique is applicable to interfaces with visual displays as well.

Progressive Prompting On Error. A conversational interface should give users the freedom to speak with minimal constraints, but yet provide assistance when needed. Yankelovich, Levow, and Marx (1995) have proposed that progressive prompting, whereby users are given increasingly directive prompts, can decrease user frustration and increase goal success. The progressive prompting in the Conversation Machine consists of suggesting strategies that we have observed people use successfully. For example, the advice to use "short, simple requests" described above was based on the observation that users who employed this strategy were often successful (Wolf et al., 1997). The highest level of prompting gives examples of requests the system can handle. Similarly, the use of examples at the highest level of prompting is intended to encourage users to switch to a goal that the system can handle (a strategy we observed in our user studies), as well as to give explicit models for a successful speaking style.

Emphasis On Frequent Transactions. As we developed the Conversation Machine to accommodate the full range of transactions handled by our banking partner's IVR, we found we needed to make some tradeoffs. As the number of utterances to be recognized increased, accuracy dropped. Consequently, to ensure adequate overall transaction accuracy we focused on accommodating a large number of ways of accomplishing frequent transactions (e.g., getting balances, making transfers), but fewer ways of accomplishing infrequent transactions (e.g., reporting a stolen credit card).

Other aspects of the system designed with the goal of facilitating error recovery and avoidance include: 1) an introduction that includes examples of utterances that work and speaking hints; 2) a help command that responds with more detailed assistance; 3) a pause function that can be used to reduce errors due to speech and noise not intended for the system; and 4) a performance aid that includes examples.

5.5   Lessons Learned

Our experience designing the Conversation Machine has taught us that conversation with a computer requires much deliberate attention to keeping the conversation on track. The system will make errors, often in rather non-human-like ways, and users may have little clue as to how to avoid or recover from errors. While the principles of human-human communication apply (Clark and Brennan, 1991), strategies for error recovery and avoidance must be designed with the limitations of the technology in mind.

6.   LOOKING INTO THE FUTURE: PATHS/ISSUES TO EXPLORE

Sitting in 1998 writing about the development of systems from "way back" in 1993 on up to 1997, we can see that there have been considerable improvements in the technology for both speech-to-text recognition and for text-to-speech output. Continuous speech is replacing isolated, systems differentiate between text and commands in dictation, speaker independence is closer to reality, and more natural sounding output is being produced incrementally. Problems remain with the 1998 generation of large vocabulary desktop dictation products (see Karat, Halverson, Horn and Karat, (in press) for a comparison of three such products). However, it is not very far out to imagine dictating an e-mail note in English and having my system translate it and read it back to me in French. We can now move beyond having to exaggerate the usefulness of speech in the interface to focusing on how such technology (which now simply works much better than it used to) can become as much a part of our systems as pointing devices and keyboards.

There are clearly many areas that need additional work. Work to explore concurrency in multi-modal interfaces would be valuable. Currently, we are limited by tools (e.g., limitations of standard system components in accepting multiple input sources) in how easy it is to explore how different modes might be integrated so that people who can talk, type, and point at the same time can do so effectively in interacting with systems. We can imagine how pointing combined with speech might have helped solve some of our design problems, but have not yet seen genera-purpose systems that do the integration well.

Do we see problems that may keep speech interfaces from becoming much more widely available in the future? Not really. Certainly speech has some disadvantages (e.g., noisiness) for some situations, but so does keyboard (e.g., requires space) or pointing (e.g., relies on displays). There are some indications that speech input may interfere with other parallel cognitive activities more than typing, but at this point it seems reasonable to expect that such impacts are minor. Would this go away with practice? We suspect it would, once we have had the opportunity to spend as much effort to tune SUIs as we have had to develop GUI designs.

REFERENCES

Clark, H. H., and Brennan, S. E. (1991). Grounding in communication. In J. Levine, L. B. Resnick, and S. D. Behrand (Eds.), *Shared cognition: Thinking as social practice*. Washington: APA Books.

Danis, C., Comerford, L., Janke, E., Davies, K., DeVries, J., and Bertrand, A. (1994). StoryWriter: A speech oriented editor. In C. Plaisant (Ed.), *Human factors in computing systems - CHI'94 conference companion* (pp. 277-278). New York: ACM.

Danis, C., and Karat, J. (1995). Technology-driven design of speech recognition systems. In G. Olson and S. Schuon (Eds.), *Symposium on designing interactive systems* (pp. 17-24). New York: ACM.

Hagberg, M., and Rempel, D. (1997). Work-related disorders and the operation of computer VDT's. In M. Helander, T.K. Landauer, and P. Prabhu (Eds.), *Handbook of human-computer interaction* (pp. 1414-1429). Amsterdam: Elsevier.

Karat, C-M., Halverson, C., Horn, D., and Karat, J. (in press). Patterns of entry and correction in large vocabulary continuous speech recognition systems. To appear in M. Altom, M. Williams, K. Ehrlich, and W. Newman (Eds.), *Human Factors in Computing Systems - CHI'99 Conference Proceedings.* New York: ACM.

Karat, J. (1995). Scenario use in the design of a speech recognition system. In J. Carroll (Ed.), *Scenario-based design* (pp. 109-134). Wiley: New York.

Lai, J., and Vergo, J. (1997). MedSpeak: Report creation with continuous speech recognition. In S. Pemberton (Ed.), *Human Factors in Computing Systems - CHI'97 Conference Proceedings* (pp. 431-438). New York: ACM.

Ly, E., and Schmandt, C. (1994). *Chatter: A conversational learning speech interface.* Paper presented at the AAAI Spring Symposium on Intelligent Multi-Media Multi-Modal Systems.

Marx, M., and Schmandt, C. (1996). MailCall: Message presentation and navigation in a nonvisual environment. In M. Tauber, V. Bellotti, R. Jeffries, J. D. Mackinlay, and J. Nielsen (Eds.), *Human Factors in Computing Systems - CHI'96 Conference Proceedings* (pp. 165-172). New York: ACM.

Oviatt, S. (1995). Predicting spoken disfluencies during human-computer interaction. *Computer Speech and Language, 9*, 19-35.

Wolf, C. G., Kassler, M., Zadrozny, W., and Opyrchal, L. (1997). Talking to the Conversation Machine: An empirical study. In S. Howard, J. Hammond, and G. Lindgaard (Eds.), *Human-computer Interaction - INTERACT'97* (pp. 461-468). London: Chapman and Hall.

Wolf, C., and Zadrozny, W. (1998). Evolution of the Conversation Machine: A case study of bringing advanced technology to the marketplace. In C. M. Karat, A. Lund, J. Coutaz, and J. Karat (Eds.), *Human Factors in Computing Systems - CHI'98 Conference Proceedings* (pp. 488-495). New York: ACM.

Yankelovich, N., Levow, G. A., and Marx, M. (1995). Designing SpeechActs: Issues in speech user interfaces. In I. R. Katz, R. Mack, and L. Marks (Eds.), *Human Factors in Computing Systems - CHI'95 Conference Proceedings* (pp. 369-376). New York: ACM.

Zadrozny, W. (1995). From utterances to situations: Parsing with constructions in small domains. *Language, Logic and Computation: The 1994 Moraga Proceedings.* Stanford: CLSI.

Zadrozny, W. (1996). Natural language processing: Structure and complexity. *Proceedings of the Eighth International Conference on Software Engineering and Knowledge Engineering* (pp. 595-602).

Zadrozny, W. , Wolf, C., Kambhatla, N., and Ye, Y. (1998). Conversation machines for transaction processing. *Proceedings of IAAI-98* (pp. 1160-1167). Menlo Park: AAAI Press/MIT Press.

2 SPOKEN NATURAL LANGUAGE DIALOGUE SYSTEMS: USER INTERFACE ISSUES FOR THE FUTURE

Susan J. Boyce

AT&T

Key words: Natural Language Understanding (NLU), Automatic Speech Recognition (ASR), Dialogue Design

Abstract: *Technology advances in automatic speech recognition (ASR) and natural language understanding (NLU) in recent years have brought us closer to achieving the goal of communicating with machines via unconstrained, natural speech. Until recently, most applications of speech recognition technology required that the user know a restricted set of command words. In contrast, the research system that is described in this chapter can understand and act upon fluently spoken language. This markedly changes the nature of the dialogue between the human and the computer. The objective of this research is to evaluate user interface design alternatives for these new natural spoken dialogues between humans and machines. In this chapter the focus is on a particular experimental vehicle, that of automatically routing telephone calls based on a user's fluently spoken answer to the question "How may I help you?". This chapter summarizes results from several studies conducted to determine how best to design the user interface for a spoken natural dialogue system.*

1. INTRODUCTION

Present-day commercially available telephony automatic speech recognition (ASR) systems offer some advantages over touch-tone systems, but also come with some limitations of their own. Most ASR systems work best when the number of active keywords is restricted to a small set, typically under 100 words. These systems vary, but for the most part they are only partially tolerant of out-of-vocabulary speech, which means that users of these systems have to know what to say and say only the words within the constrained grammar. In addition, some commercially available systems insist that the user wait until the prompt has ended before allowing the user to input the next command. These limitations have dictated the course of ASR user interface design for the last decade.

Because of these limitations, ASR systems typically use a menu-based approach, where all of the available commands are listed out for the user in auditory prompts. In addition, these prompts often have to include instructions about the precise format the user's speech must follow and when the user may begin to speak (Brems, Rabin, and Waggett, 1995). For these reasons, having a spoken dialogue with an ASR system can seem more like barking orders than talking to another human.

In contrast, advances in large vocabulary speech recognition combined with advances in natural language processing are bringing us closer to a time when we may be able to speak to computers in more unconstrained, natural speech. When these systems become a reality, designers will need to re-think the rules that they have followed previously for designing the user interface. Some of the lessons of the past may still be relevant, but new ones will have to be learned.

The next generation of natural language spoken dialogue systems will differ from the current generation ASR systems in a number of ways. First, users will not have to know a specific set of vocabulary words in order to speak to the system. Users will be able to state their requests in their own words and have the system map the appropriate meaning on to these words. Second, users will be able to speak much longer utterances than they can with the current technology. Third, users will be able to give multiple pieces of information in a single request. Current ASR systems often force users to give only a portion of the information needed to complete a task on a given turn, for example:

System:	Please say collect, calling card, third number, person to person or operator, now.
Caller:	*Calling card*
System:	Please enter or say your calling card number, now.
Caller:	*908 949 1111 0000*

The next generation technology will allow a caller to say "I'd like to place a calling card call. My card number is...". Finally, and perhaps most importantly, the next generation of technology should allow for much easier recovery from recognition errors due to the flexibility of the grammars.

These changes will have broad ramifications for how best to design the user interface for voice services. No longer will it be possible to list all the system's capabilities in a menu, since the number of options will be greatly increased. Some other method will be needed to communicate the capabilities of the system to the users. However, the improvements in error handling should make the dialogues easier to design and better liked by users.

This chapter focuses on a series of experiments that were conducted at AT&T. The research team[1] was interested in investigating next generation natural language capabilities. The purpose of the studies summarized here was to begin to examine the user interface issues that arise when designing for the next generation of technology. The system the research team developed was an experimental spoken natural dialogue system designed to handle routine calls that come in to AT&T operators.

1.1   The Call Routing Task

The call routing task that was studied involved classifying users' responses to the open ended prompt "*How may I help you?*" from a telephone operator (Gorin, 1995). The goal of this experimental system is to classify responses as one of 17 call types so that the call can then be routed to an appropriate destination. For example, if a person said "*Can I reverse the charges?*", the appropriate action is to connect them to an automated subsystem which processes collect calls. If the request was "*I can't understand my bill*", the call should be routed to the appropriate customer service agent.

1.2   Design Process

In order to produce a natural dialogue design, an iterative process of design and testing was used (Day and Boyce, 1993). The first stage of design was to collect and analyze human-to-human dialogues for the call routing task (see Gorin, Parker, Sachs, and Wilpon, 1996). These were conversations between callers and live operators. In the second phase, the initial design of the human-machine dialogue was analyzed and the elements of the dialogue that could make the human-computer dialogue seem unnatural were identified. The third phase was to conduct Wizard of Oz experiments as defined in Gould, Conti, and Hovanyecz (1983) to investigate how best to design the interface when modeling the human-human dialogues was impossible.

  With this technique the speech recognition and natural language understanding components of the system are simulated, although the user does not need to know this. The user calls in to the system and is greeted by the automated system. The experimenter monitors the call, and it is the experimenter, not the system, that determines how to respond to the caller. The experimenter can "simulate" an error or a correct response by pushing the appropriate key on a computer that is controlling which system prompts get played back over the telephone to the caller. This kind of experiment can be very valuable for evaluating user interface components, particularly error recovery strategies, since the experimenter can tightly control when and where "errors" occur.

1.3   Analysis of Human-to-Human Dialogues

The first phase involved gaining a better understanding of how callers express their requests to humans and how most human agents elicit clarifying information (Gorin et al., 1996). By doing this, important data for algorithm and technology development was collected in addition to important information for the design of the user interface. By closely matching the wording of the system prompts to the words used by the human we thought that a greater degree of naturalness could be achieved.

2.   ANTHROPOMORPHISM AND USER EXPECTATIONS

The first design question that arose as a result of the new technological capabilities was whether or not the automated system should remain a distant, formal machine that speaks in the third person, or whether the automated system should speak in a more human-like style. There has long been a prohibition against making computer systems "sound" human when they communicate (Shneiderman, 1992). The primary objection to anthropomorphic systems has been that they create unrealistic user expectations that lead to errors and then disappointment with the system. Some older human factors research indicated that this was true and, additionally, that there can be negative reactions to a user interface that is too cutesy or talkative.

However, some of the data collected on this issue has been with screen-based systems for which the user types input on a keyboard. It is possible that since the capabilities of natural spoken dialogue systems more closely match the capabilities of humans that the negatives associated with anthropomorphic interfaces are moderated. In addition, it is possible that users have been exposed to a much wider variety of automated services since these studies were conducted and that this exposure has resulted in a change in user perception about anthropomorphism.

Therefore, the first experiment that was conducted was designed to determine what aspects of a system might make the system seem more human-like and to determine if making the system seem more human-like had a positive or negative effect on users' satisfaction with the system.

Anthropomorphism is often defined as having the system refer to itself as "I", as in "How may I help you?". However, for the purposes of this study a second dimension of anthropomorphism was considered; whether or not the language used for the prompts was casual or formal. Typically, prompts for interactive voice response (IVR) systems are written in complete sentences with vocabulary that is well-understood but tends to be formal in tone. In contrast, live operators assisting callers on the phone often phrased their requests in partial sentences, even single words, and the intonation carried much of the meaning. For example, if a caller asked for assistance placing a long distance call, an operator might ask "The phone number?" to prompt the caller to say the number they were trying to reach. A typical automated system would be more likely to say something like "Please enter or say the area code and telephone number you wish to call now".

There are obvious reasons for these differences in style. The human operator has the ability to detect when the caller is confused and misunderstands the question, and the operator can offer more help before the dialogue gets hopelessly confused. If a caller hesitates for a moment, the operator may say "May I have the phone number you are trying to call?" as a way to further clarify what information is needed. Similarly, if the caller doesn't provide the area code, but it's required, the operator can easily determine which information is missing and ask just for the missing pieces. In contrast, when designers write prompts for IVR systems, it is their fervent hope that most callers will understand what is requested from them on the first try, since error recovery is so long, tedious, and often ineffective.

2.1   Anthropomorphism Experiment

This study tested users' reactions to an automated natural language system that used prompts and dialogues modeled on actual human operators' words. The four conditions of this experiment are shown in the example dialogues in Table 1. The two dimensions studied were whether or not the system referred to itself as "I" and the "Casual/Formal" dimension.

Table 1.   An example dialogue (request for an area code) for each of the prompt styles tested.

	Casual	Formal
"I"	S: How may I help you? U: I need the area code for Morristown, New Jersey. S: Okay, you need me to give you an area code. U: [either silence or] Yes. S: City and State? U: Morristown, New Jersey. S: The area code is 201. S: Anything else I can help you with? U: No, thanks. S: Thank you for using AT&T.	S: Please tell me how I can help you, now. U: I need the area code for Morristown, New Jersey. S: Do you need me to give you an area code? U: Yes. S: Please say the city and state for which you would like the area code, now. U: Morristown, New Jersey. S: The area code is 201. S: If there is anything else I can help you with please tell me now. U: No, thanks. S: Thank you for using AT&T.
No "I"	S: What would you like? U: I need the area code for Morristown, New Jersey. S: Okay, you need to know an area code. U: [either silence or] Yes. S: City and State? U: Morristown, New Jersey. S: The area code is 201. S: Is there anything else you need? U: No, thanks. S: Thank you for using AT&T.	S: Please state your request, now U: I need the area code for Morristown, New Jersey. S: Do you need to know an area code? U: Yes. S: Please say the city and state for which you would like the area code, now. U: Morristown, New Jersey. S: The area code is 201. S: If you have an additional request, please say it now. U: No, thanks. S: Thank you for using AT&T.

* System is abbreviated with S and User is abbreviated with U.

The prompts for the Casual/"I" condition were modeled after the kinds of questions live operators asked of callers. The prompts for the Formal/No "I" condition were written following guidelines and rules of thumb that currently exist for writing prompts for IVR systems. Casual/ No "I" and Formal/"I" conditions were designed to help determine whether or not people simply object to a machine referring to itself as if it were human (i.e., anthropomorphism) separately from whether they prefer a more natural conversational style or a traditional formal IVR prompt style.

As in the previous experiment, callers interacted with a Wizard of Oz simulation of a real system. With this method, the error rate that subjects experienced could be controlled; the same error rate occurred for each of the versions of the system tested. Each version of the system was tested with 21 users (84 subjects overall). Users placed seven calls to the system and then completed a questionnaire to measure their satisfaction with the system. Each subject was interviewed to gain additional feedback about the system.

The ratings of overall satisfaction, shown in Table 2, indicated that users were significantly more satisfied with the "I" conditions than with the No "I" conditions (p<.01). There was no effect on overall satisfaction ratings of the Casualness manipulation. The latter of these results was somewhat surprising in that the differences between the Casual and Formal versions of the system seemed much more pronounced than the differences between the "I" and No "I" conditions. Formal prompts were longer than the Casual prompts and this seemed like a salient difference. It took, on average, almost 8 seconds longer to complete a task with the Formal prompts than it took with the Casual prompts. Nonetheless, the primary finding for the ratings of overall satisfaction was that users preferred the systems that used the personal pronoun "I".

Table 2.   Ratings of overall satisfaction with each version of the system.

Survey Question: Overall, how satisfied or dissatisfied are you with this operator service. (1=Very Satisfied, 6= Very Dissatisfied)

	Casual	Formal	
"I"	1.81	1.91	1.86
No "I"	2.24	2.43	2.34
	2.03	2.17	

During the interview following the completion of the questionnaire users were asked whether or not they noticed that the system referred to itself as "I". Eighty percent of users that had experienced an "I" condition did not notice the use of the personal pronoun. When they were asked whether they liked or disliked the use of "I," all users answered that they either liked it or that it didn't make a difference to them. Hence it seemed that the effect of "I" was positive but also fairly subtle, in that it wasn't very noticeable to users.

Another interesting finding was that users' ratings of how satisfied they were with the speed of the interaction seemed more correlated with their overall satisfaction with the system and very unrelated to the amount of time the interaction actually took (see Table 3. The transaction times appear in the parentheses). In other words, time flies when you're having fun.   This finding is in direct contradiction to most human factors guidelines which recommend that prompts should be as brief and as succinct as possible. What these data seem to argue is that it isn't the brevity of the experience that matters so much, but rather how much the user *likes* the experience.

Table 3. Ratings of satisfaction with the speed of the systems with actual
transaction times in parentheses.

Survey Question: How satisfied or dissatisfied are you with the speed of the operator
service. (1=Very Satisfied, 6= Very Dissatisfied)

	Casual	Formal	
"I"	1.76 (43.2 s)	1.67 (51.3 s)	1.72
No "I"	2.10 (43.7 s)	2.33 (50.8 s)	2.22
	1.93	2.00	

However, participants in this study were not as easily duped when they were
asked about their satisfaction with the error rates for the systems. Speech
recognition errors were simulated in this Wizard of Oz study, occurring randomly
throughout the user's experience. The error rate was held constant at 85% correct
recognition across the seven tasks for each system. No significant differences were
detected in participants' ratings of satisfaction with the system performance as can
be seen in Table 4. This seems to suggest that although users may be inaccurate in
their estimates of the length of time a task takes, they are more sensitive to the
number of mistakes the system makes. In other words, a system with an appealing
dialogue style may lead users to pay less attention to how long a task takes to
perform, but it probably won't be successful at getting users to overlook the number
of recognition errors the system makes.

Table 4. Ratings of satisfaction with system performance.

Survey Question: How satisfied or dissatisfied are you with the number of errors that the
operator service made. (1=Very Satisfied, 6= Very Dissatisfied)

	Casual	Formal	
"I"	2.43	2.57	1.72
No "I"	2.38	2.62	2.22
	2.41	2.59	

In summary, users were more satisfied with systems that referred to themselves
as "I" and there was no cost associated with longer prompts (and no gain with
making the prompts more casual). These results should not be interpreted as license
to make prompts wordy or "chatty". When writing these anthropomorphic versions
of this system, every attempt was made to keep the prompts from sounding
"cutesy", overly friendly, funny, or chatty just for the sake of it.   The
anthropomorphic versions were written by modeling the speech of human operators.
No doubt, one could very easily take a different approach and write
anthropomorphic prompts that are disliked by users and would very negatively
impact customers' willingness to use the system. However, if anthropomorphism is
used judiciously in a system designed for occasional use, it can enhance users'
satisfaction with the system, often without the user becoming overly aware of the
anthropomorphism.

3.   ISSUES FOR NATURAL DIALOGUE DESIGN

The previous study indicated that modeling the automated dialogue from the live operators' speech resulted in the highest satisfaction. However, not all aspects of human-computer dialogue can be modeled after human-to-human dialogue. Some elements of the human-computer dialogue are necessary simply because the automated system does not have all of the capabilities of a human listener.

In designing the automated call routing system, there were several aspects of the dialogue for which there was not a ready analog available in human to human conversation because the machine was prone to making mistakes that a human very rarely would make.   These dialogue elements include the initial greeting, confirmation of the user's request, disambiguation of an utterance, reprompts, and knowing when to bail out to a human to complete the transaction.

3.1   Initial Greeting

A frequently-voiced concern about designing very natural human-computer dialogues is that early in the interaction, users are likely to assume that the system has greater capabilities than it actually has, and therefore attempt to speak in a manner that the system has little probability of understanding. Designing the right initial greeting is necessary to appropriately set user expectations.

3.2   Confirmations

Since any spoken dialogue system will not perfectly interpret the user's speech all of the time, a necessary element of a human-computer dialogue is a confirmation step. During confirmation, the system repeats what it thinks was said, giving the user a chance to confirm or deny the system's interpretation. The confirmation strategy depends on the confidence the system has in its interpretation. For example,

System:	How may I help you?
User:	*Yes, could you give me the area code for Morristown, NJ?*
System:	Do you need area code information?
User:	*Yes, I do.*

A human is not likely to say the phrase "*Do you need area code information?*" in such a context (Clark and Schaefer, 1987). The issue for confirmations is to ask the question in a way that more closely resembles the way a human might ask for confirmation, in order to elicit the appropriate user response.

3.3   Disambiguating an Utterance

In some cases, the automated system is going to come up with more than one interpretation of the user's speech. There are several ways that the system could ask for clarifying information from the user. One straightforward way to do this would be to simply ask the user, "*Do you want A or B?*".Another way might be to ask a yes/no question "*Do you want A?*" so that if the answer is no then it might boost the probability that B is the correct choice. Which strategy is most appropriate will depend on the relative confidences of likely interpretations and which is most effective at completing the dialogue quickly and naturally.

3.4   Reprompts

A low confidence from the natural language processor indicates that it didn't understand the user's utterance. In these cases, rather than asking for confirmation of information that is almost certainly wrong, it is better simply to ask for the user to repeat his or her request. This step is referred to as a reprompt. Often in human-computer dialogs this step takes the form of "*Sorry, please repeat*". The system admits culpability, then as quickly as possible asks for the information to be repeated. This phrasing does not however, provide any information on what went wrong.

In contrast, human listeners have a wider repertoire of responses available to them to communicate to the speaker which elements they didn't understand in an utterance. Possibilities are that they didn't hear properly, or that they heard the utterance but didn't understand it, or that they heard and partially understood, but need more information. All of these states can be quickly communicated between humans using pauses, prosody and the content of the response. A challenge for human-computer dialogues is to intelligently mimic these devices in order to communicate to the user how the conversation has failed so that the user can provide useful inputs.

3.5   Turntaking

Humans take turns very quickly and effectively under normal circumstances in human-to-human conversation. However, current technological limits in recognition speed and also in difficulty determining when a speaker has finished cause awkward pauses in human-machine dialogues. The design challenge is to develop a system that can accurately detect when the speaker is finished and quickly move on to the next step in the dialogue.

3.6   When to Bail Out

Sometimes a human-computer dialogue experiences repeated breakdowns, as shown by either low system confidence or repeated responses of "no" to confirmation prompts. In such cases, a human should be brought in to complete the transaction. The design question is how to decide when a situation calls for human intervention to ward off undue user frustration. The correct criterion might be number of recognition errors in a row, or it might be the number of errors that have occurred overall in the dialogue. It seems likely that the right answer to this question may very well depend on characteristics of the user population and of the task being accomplished by the automated system. The goal should be to bail out before the user's frustration causes a negative perception of the system.

No doubt there are many other design issues that deserve some study when developing spoken natural dialogue systems. This list represents the issues that the research team at AT&T decided to study in more detail because they were critical for our application. The remainder of this chapter focuses on the issues of initial greetings, confirmations, reprompts, and turntaking.

4. ESTABLISHING USER EXPECTATIONS IN THE INITIAL GREETING

One of the most difficult problems in designing a spoken natural dialogue system is establishing in the user's mind what the system can and cannot understand.   The basic problem is this: if the system sounds too much like a human, so that it is virtually indistinguishable from a human, then the user can reasonably expect that the system will understand like a human, a feat that machines are not up to yet. On the other extreme are systems that prompt the user with a list of command words that the system knows: a menu. In this case, the user probably has the expectation that this limited set of words is all that the system understands, and users must constrain themselves to this set, whether or not the set meets the user's needs in the dialogue. The capabilities of the "How May I Help You (HMIHY) System" fall somewhere in between these two extremes. The list of possible "commands" is too long to be presented in a menu so a more open-ended prompt, such as "How may I help you?" seems appropriate. But, from listening to many thousands of calls to live operators, it became obvious that if callers believe they are speaking to a human operator, their requests are often very long and complicated. Therefore the design team wanted the spoken natural dialogue system not be mistaken for a human operator since it was unreasonable to expect that the HMIHY system would handle very long complicated requests as well as a human operator.

Hence, the goal of the first user study was to come up with an initial greeting that let the caller know that they were talking to a machine, not a human. The hope was that if users knew they were talking to a machine, they would bring what they know about machines to bear on the dialogue. This should have the effect of altering their requests in some way, and perhaps it would make them shorter (Falzon, 1990; Franzke and Marx, 1993; Kennedy and Wilkes, 1988; Richards and Underwood, 1984).

Initially, two ways in which to communicate to callers that they were speaking to a machine were tested. The first was to explicitly say it was a machine in the first announcement heard by the caller; the second was to use Text-To-Speech (TTS) synthesized speech to make the system "sound" like a machine. These alternatives were tested using the Wizard of Oz method with a group of callers to see if changes to the initial greeting could affect the length of the utterances callers might use to state their requests.

4.1   Initial Greeting Experiment

The initial greetings that were tested can be seen in Table 5.

Table 5.   Prompts tested in the Initial Greeting Experiment.

Condition	Prompt
Standard	AT&T. How may I help you?
Short Automated	AT&T Automated Customer Service. How may I help you?
Long Automated	AT&T Automated Customer Service. This system listens to your speech and sends your call to the appropriate operator. How may I help you?
Long Automated-TTS	AT&T Automated Customer Service. This system listens to your speech and sends your call to the appropriate operator. How may I help you?

Ultimately, four greetings were chosen for the test. The first greeting, labeled Standard, is the greeting that was used in the AT&T network at the time of this test by the human operators. In the network, this greeting is pre-recorded and is played across the phone line as soon as the operator answers a call.   The Standard condition mimics the situation that a caller experiences when calling in to a live operator. The same voice used in the AT&T network at the time was used for the Standard prompt (and the other recorded human speech prompts).

The Short Automated prompt differed from the Standard greeting in that it included the phrase "automated customer service". It was thought that this might be sufficient to cue callers that they were speaking to a machine and not a live operator. The Long Automated prompt took this a step further. Not only does it include the "automated customer service" prompt but also a lengthy description of what the system is attempting to do[2]. No human would say a phrase like this, so it was thought that this might be an additional cue to the caller that the system was a machine. Users might miss a single word like "automated" in a fast prompt. The long prompt was as explicit and obvious as possible to try to remedy this.   Finally, this same long prompt was used for the text of the synthesized speech prompt. The TTS was highly intelligible but clearly not a human voice.

A total of 545 callers experienced some version of the test system. Callers did not know they were speaking to a Wizard of Oz system. After their requests were handled by a combination of the HMIHY system and live operator, callers were asked to rate the automated portion of the call as to whether it was Excellent, Good, Fair or Poor. The callers' utterances were transcribed and the number of words they used to state their requests was counted for each condition. The results are seen in Table 6.

The standard greeting produced the longest utterances on average (12.99 words) and the Short Automated greeting did little to shorten the length of callers' utterances (12.43 words). These means are not significantly different. However, the Long Automated prompt did produce reduced average utterance length as did the TTS version of this prompt. Each was significantly different from the Standard greeting ($p < .05$) and from each other ($p < .05$). Hence, it seems that callers pretty much ignored the phrase "Automated Customer Service". Only when the long prompt was used did the callers modify their utterances. The use of synthesized speech in conjunction with the long prompt was the most effective at getting the utterance length reduced. However, results from the customer satisfaction measures indicated that callers disliked the synthesized speech so much as to offset any gain in shorter utterances that the TTS offered over the Long Automated prompt alone.

Table 6.   Average number of words callers used to state their request as a function of the initial greeting.

Prompt	Avg. No. of Words in Request	N
AT&T. How may I help you?	12.99	121
AT&T Automated Customer Service. How may I help you?	12.43	143
AT&T Automated Customer Service. This system listens to your speech and sends your call to the appropriate operator. How may I help you?	10.52	145
AT&T Automated Customer Service. This system listens to your speech and sends your call to the appropriate operator. How may I help you? [TTS]	8.47	136

This study indicated that the best choice of initial greeting was the Long Automated prompt. There could be at least two explanations for why this was the case. It could be that the wording of this prompt makes it clear that it is an automated system. The second possibility is that the wording wasn't as important as the length of the prompt. That is, it could be that the length of this prompt gave callers sufficient time to re-think and re-word their request.

A subsequent study was conducted to further refine the wording of this prompt and to test another alternative: playing an "audio logo" or sound effect at the beginning of the greeting as a way to cue the caller that it is an automated system. This proved to be very effective, achieving results that were akin to those achieved with the Long Automated prompt, but requiring much less time since the shorter version "How may I help you" could then be used.

The content of the first prompt is an important element in setting user expectation. By making the first prompt explicitly state that the speaker is a machine and not a human, callers stated requests with shorter utterances that were easier for the HMIHY system to understand. Not only the content of the prompt mattered, but also the voice (TTS or recorded human speech), and sound effects were effective for cueing callers that they were talking to a machine.

This look at the initial greeting of the system only brushed the surface of setting user expectations. Most of the dialogues require several turns taken by the caller and system to complete the task, and at each stage the caller needs to understand what is and isn't allowed vocabulary. Further work remains to be done to explore more completely users' mental models of the systems to which they are speaking, since this determines what they will and will not choose to say.

5.   IDENTIFYING RECOGNITION ERRORS THROUGH CONFIRMATIONS

As mentioned earlier, human listeners have an amazing ability to detect and correct misunderstandings in conversations with other humans. Properly detecting and correcting errors in human-machine dialogues is awkward and tedious under most circumstances. One method machines commonly use to detect errors in recognition is to use a confirmation step in the dialogue. In the natural dialogue system, when a user has responded to an open-ended query with a relatively long reply, it was often necessary for the system to confirm with the user that its interpretation of the user's speech is correct, for example:

System:	How may I help you?
Caller:	*I was trying to place a call and must have dialed a wrong number, can I get credit for that?*
System:	Do you need me to give you credit?
Caller:	*Yes.*

These kinds of confirmations rarely occur in natural human-human conversations in the form that is often required in human-machine dialogues (Clark and Schaefer, 1987). This poses a problem when we try to model human-machine dialogues after human-human dialogues. An early study that was conducted prior to the work on the HMIHY system evaluated various strategies for confirmation of digit strings in ASR systems in order to determine the most preferred method. The purpose of this study was to find alternatives to the traditional style of confirmation that are more like human conversation and are shorter and more efficient than the traditional method.

5.1   Confirming digit strings in spoken dialogue systems

This study was conducted within the context of developing the user interface for a voice dialing system. Callers were permitted to say a telephone number they wished to call. If the system was unsure about one or some of the digits in the telephone number, then some confirmation step was necessary in the dialogue. Three alternatives for confirmation were tested. The first was the standard method that was used at the time: simply having the system say "Did you say..." followed by the system's interpretation of the digit string. Users would then respond "yes" or "no" to this question. If the answer was "no" then the system would re-prompt for the digit string again, a tedious process that always met with user frustration.

A second strategy that was evaluated was an attempt to mimic the "readback" strategy often used in human-human conversation. With this method, the listener repeats back to the speaker the digits with a flat intonation. The expectation is that as the speakers listen to their digits being read back, and if there is an error they will jump in to correct it. This is a strategy often used by sales representatives who take credit card numbers over the phone, for example:

Sales Rep:	Can I have your credit card number please?
Caller:	*1234 5678 2345 9911*
Sales Rep:	That's 1234 5678 2345 9991
Caller:	*No, 9911.*

This was called the "Just Say No" strategy. An earlier study using this strategy in a human-machine dialogue indicated that users were not likely to interrupt the machine when it made an error reading back the digits, unless they had been explicitly told that they could do so.

The third strategy tested was a modification of this where the confirmation question is carried by the inflection in the voice of the system repeating back the digits (the "Question Inflection" strategy). The use of rising inflection at the end of the digit string was used to indicate that the caller was supposed to answer yes or no. For example, the system might say in response to a caller "908 949 4651?" where the inflection on the last digit "1" indicates that it is a question.

A total of 24 participants placed four telephone calls each with each of the three different versions of the simulated voice dialer. After experiencing each system, a questionnaire was administered to get the user's impressions of the system. In addition, the amount of time it took to actually complete the call was measured for each of the systems.

Transaction times were significantly shorter for the question inflection strategy than for the other two. Obviously, the question inflection strategy was shorter than the "Did you say..." strategy simply because fewer words were needed to elicit a response from the caller. The question inflection strategy was shorter than the "Just Say No" due to the length of time required to give the callers the instructions in the prompts that they could interrupt the system if it made an error. Given that these strategies differed from one another in length by design, the real point was to evaluate whether or not the new strategies were as effective as the traditional confirmation strategy.

Users were asked to specify the strategy that they liked the best and the strategy that they liked the least. Figure 1 shows the percentage of participants who most preferred each of the strategies.

As you can see from these data, the least preferred system is the "Did you say.." which is in wide use in ASR systems today. Most users found this method to be the most tedious. The most frequent complaint was that it took too long.

These data indicated that, when a confirmation step is necessary, the quickest and most preferred method to use is the question inflection on the final digit of the string. This can be somewhat tricky to implement since many IVRs use concatenated recorded speech to do the digit confirmations. Using the question inflection requires that the system store an additional final question pronunciation for each of the possible digits. However, if the system is frequently used by the same caller, it is worthwhile to invest time and effort to make this step as painless as possible as this is a part of the dialogue that generates a large amount of user dissatisfaction.

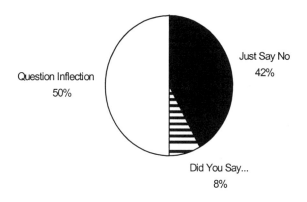

Question Inflection
50%

Just Say No
42%

Did You Say...
8%

Figure 1. Ratings of the most preferred method of confirming digit strings.

5.2 Confirmation of Topic in a Spoken Natural Dialogue System

In an early version of the HMIHY system, it was often necessary to confirm with a caller that the system had correctly identified the topic of the conversation before proceeding with the dialogue. For example, if a caller said, "I'd like to make a call and reverse the charges", the system might need to confirm with "Do you want to make a collect call?" before asking the next question. However, this wording bears a marked resemblance to the "Did you say" wording that the previous study indicated that users did not like, so an alternative was required. The alternative was to model the "Just Say No" strategy from the previous experiment. That is, have the system respond with "You want to make a collect call" as a declarative statement as the confirmation. Even though the previous experiments indicated that callers sometimes would not correct an error, the hope was that with a very flexible, human-sounding natural dialogue this problem might disappear.

To test this, callers were given either the "Do you want to make a collect call?" form, which was called the Explicit confirmation because an explicit yes or no was required from the caller before proceeding, or the "You want to make a collect call." version, which was termed the Implicit confirmation, because it did not require any answer from the caller.

Explicit Confirmation example:

System: How may I help you?
Caller: *I'm trying to find a number for Robert Smith in Tacoma Washington.*
System: Do you need me to give you directory assistance?
Caller: *Yes.*

Implicit Confirmation example:

System: How may I help you?
Caller: *What's the area code for Allentown Pennsylvania?*
System: You need area code information.
Caller: *Yes.*

For 259 callers to the simulation, the explicit confirmation strategy was used; 145 were given correct feedback, 114 were given incorrect feedback (to simulate classification errors).   A separate sample of 233 callers heard the implicit confirmation strategy; 119 received correct feedback and 114 were given incorrect feedback.   The strategies were evaluated by comparing the probability of successfully repairing the dialogue from a simulated error.

Both strategies were very successful if the machine had correctly interpreted the user's speech. However, users were somewhat less successful at repairing errors in the dialogue when the implicit confirmation was used as compared to the explicit confirmation (see Table 7).

Table 7.   Callers responses to Explicit and Implicit Confirmations.

	Explicit (Do you want X?)	Implicit (You want X.)
"Yes..." for Correct Feedback	91%	93%
"No..." for Incorrect Feedback	74%	63%

For trials in which a recognition error was simulated, only 63% of the users corrected the error by saying "No," then optionally following this with the correct information. Furthermore, 15% of the callers with Implicit Confirmation failed to correct the error at all, as compared with only 6% of calls that received Explicit Confirmation (see Table 8). It is possible that these users were unclear about how to interrupt the system to correct the error. These data suggest that the explicit confirmation method, albeit possibly less natural, is the more robust strategy to pursue given that more of the errors get correctly repaired. However, it is possible that with more experience callers would learn how to use the implicit confirmation method and would prefer it.

Table 8.   Responses to incorrect confirmations classified into categories based upon semantic content of response.

Caller Response	Explicit (Do you want X?)	Implicit (You want X.)
Failed to correct error (silence)	6%	15%
No (or synonym)	43%	39%
No + Correct Information	31%	23%
Just Correct Information	14%	15%
Request to repeat confirmation (i.e. excuse me?)	6%	8%

Another way the strategies were evaluated was to look more closely at the speech that callers used to respond to the confirmation. To maximize speech recognition performance, a good confirmation strategy is one in which most or all users know what to say to respond and, furthermore, that they say only what is required to move on to the next step in the dialogue and not volunteer a lot of extraneous information.

To evaluate these methods of confirmation for extraneous speech, the callers' responses to the confirmation were classified into the categories listed in Tables 8 and 9. Table 8 the percentage of each response type when a classification error was simulated, and Table 9 shows the responses when the confirmation was correct.

Table 9.   Responses to correct confirmations classified into categories based upon the semantic content of the response.

Caller Response	Explicit (Do you want X?)	Implicit (You want X.)
Yes (or a synonym)	67%	75%
Yes + Short Phrase (i.e. Yes I do, please)	18%	16%
Yes + New Information	6%	2%
Only New or Clarifying Info	6%	5%
Request to repeat confirmation (e.g. "excuse me?")	4%	2%

In the case where the confirmation was correct, most of caller responses were either a simple "yes" or "yes" followed by a small handful of easily recognized carrier phrases. However, the remaining 15% of callers for the explicit strategy and 9% for the implicit strategy added significantly more information to their response than "yes" and some skipped the "yes" altogether. In 6% of the cases callers responded to an explicit confirmation like "Do you want to make a collect call?" by saying "Yes, to 908 949 1111". These are labeled Yes + New Information in the table above. Similarly, another 6% responded to this explicit confirmation by simply providing the new information and skipping the "yes".

The two strategies that were tested didn't differ significantly in the degree to which they produced extraneous speech. However, they did indicate that a more complicated speech recognition grammar is necessary at this stage in the dialogue than was previously thought. That is, more than a simple "yes/no" grammar will be necessary at this stage of the dialogue in order to adequately deal with the range of responses callers will make to confirmations. The grammar will need to include "yes" and "no" and also digits and all of the words included in the grammar for the initial greeting.

Table 8 shows the results when the caller is given incorrect feedback in the confirmation. These data indicate clearly that one very natural strategy the callers adopted was to say no followed by the correct information. Again this indicates that more complicated speech processing would be necessary during this stage in the dialogue than simply running a "yes/no + carrier phrase" grammar.

For the HMIHY system the resolution was to have the entire contents of the top-level grammar running in addition to the yes/no and the digit grammar in order to solve these problems. That is, all the words that could be understood at the "How may I help you" prompt were also active at this stage in the dialogue. This allowed for dialogues like the following to happen.

System:	How may I help you?
Caller:	*I'd like to make a call and charge it to my calling card.*
System:	Do you want to make a collect call?
Caller:	*No, a calling card call.*
System:	May I have your card number please?

This allowed for the caller to re-assert their initial topic of conversation during the response to the confirmation, which in some sense makes the confirmation seem like less of a wasted step in the dialogue. A side benefit of this was that the system was also able to tolerate complete changes in topic.

System:	How may I help you?
Caller:	*I'd like to make a call and charge it to my calling card.*
System:	Do you want to make a card call?
Caller:	*Actually, I don't have my card with me. Can I make it collect?*
System:	Do you want to make a collect call?
Caller:	Yes.

To summarize, confirmation steps are often necessary in automated dialogues in order to ensure that the system has correctly understood the user's speech. However, confirmations are often very unnatural and lead to the perception that the system is slow. In these experiments, several alternatives for confirmations for digit strings and for topic confirmation in a natural spoken dialogue system have been explored.

6.   REPAIRING RECOGNITION ERRORS WITH REPROMPTS

There are occasions with automated dialogue systems in which the system is unable to interpret the user's request at all, and knows that it does not have an interpretation. In these cases it is necessary to ask the user to restate his or her request. This stage of the dialogue is called the reprompt. Reprompts occur either in the case where the automated system has no interpretation or in the case where the automated system has offered an interpretation in a confirmation which has been rejected by the user, but no additional information has been offered by the user.

Reprompts do occur in dialogues between human speakers so there is some basis for modeling the automated system after human behavior. However, human speakers and listeners are adept at using intonation, pauses and back-channels as devices to cue the speaker as to whether the listener didn't *hear* the request or didn't *understand* the semantic content of the request. Although it would be fascinating to try to model the subtleties of this with an automated system, this was outside the scope of the project.

However, an important problem for spoken natural dialogue systems is that, when users are reprompted, they often use the exact phrase over again; the same phrase that was not recognized in the first place. This may be useful if the first try suffered from background noise or interruptions such as lip smacks or hesitations.

But, if the initial utterance was not recognized because the response was out of the machine's grammar, then a simple restatement of the utterance will not produce better recognition on the second try.

The aim with this study was to develop a reprompt phrase that produced clearer, more concise speech from the user for the second utterance than was received for the first. This would give the automated system the highest probability of correctly classifying the call on the second try.

6.1   Reprompt Experiment

The goal in this experiment was to determine whether or not users' responses to reprompts could be altered based upon the wording of the reprompt phrase. Specifically, the goal was to develop a reprompt wording that produced the most "interpretable" speech on the second utterance.   As in the previous experiments, this study was conducted with a Wizard of Oz simulation of the HMIHY system to handle call routing for operator services. This test compared several different reprompt wordings for the initial classification request. An example is below:

System:	How may I help you?
Caller:	*I'd like to make a call and charge it to my calling card.*
System:	I'm sorry. How may I help you?

The wording of the reprompt varied in terms of how instructive it became (i.e., "I'm sorry. Please briefly tell me how I may help you?"), versus how short it was (i.e.,"I'm sorry. How may I help you?"). These reprompts were evaluated in terms the semantic content of the responses given to the reprompt.

Each response was classified as being one of the following:

◆ Same – Response to the reprompt exactly the same as response to initial prompt.

◆ Rephrased – Response to the reprompt was the same request, but used different words and was about the same length as the initial request.

◆ Shorter – Response to the reprompt was the same request but stated more concisely.

◆ Fewer Ideas – Response to the reprompt contained less information than the initial request.

◆ Longer – Response to the reprompt was the same request but the caller used more words to state the request.

◆ More Ideas – Response to the reprompt contained the initial request with some additional information.

◆ Changed Request – Response to the reprompt was for a different service.

Here are some example dialogues from some of the categories.

More ideas:

System:	How may I help you?
Caller:	*I'd like to make a call and charge it to my calling card.*
System:	I'm sorry. How may I help you?
Caller:	I'd like to make a calling card call to 908 949 1111.

Changed request:

System:	How may I help you?
Caller:	*I'd like to make a call and charge it to my calling card.*
System:	I'm sorry. How may I help you?
Caller:	I want an operator.

161 callers experienced the Short ("I'm sorry. How may I help you?") Reprompt and 215 callers experienced the Instructive ("I'm sorry. Please briefly tell me how I may help you.") Reprompt. The percentage of responses that fell into each category is shown in Table 10.

Table 10. Percentage of responses to reprompts by category.

Category of Response	Short Reprompt (% of responses)	Instructive Reprompt (% or responses)
Same	42%	34%
Rephrased	6%	11%
Shorter	14%	18%
Fewer Ideas	8%	6%
Longer	14%	11%
More Ideas	12%	12%
Changed Request	4%	8%

As can be seen in the table, users as a whole did not adopt a single strategy. Some of the caller strategies in response to reprompts are likely to produce better results than others, in terms of recognition accuracy. For example, if the system had difficulty understanding the caller's initial request for the HMIHY system, it is probably not going to help for the second prompt to be longer or contain more ideas since the primary obstacle for successful recognition is that the utterance is too long. So, if the caller managed to have a shorter or rephrased request, these could be beneficial strategies in terms of improving the probability for successful dialogue. In Table 11 are listed those strategies that are considered beneficial or "productive" in terms of improving recognition performance.

Based upon this analysis, it appears that there is some advantage to the Instructive Reprompt in that 11% more of the caller responses were classified as "productive" than with the Short Reprompt.

This study demonstrated that even small changes in wording of the reprompt can have an effect on the kinds of responses callers will have to the reprompt. In this study a fairly minor change in the reprompt wording resulted in a larger percentage of productive responses. Other alternative reprompts may produce an even higher percentage of productive speech in the second utterance and this is an area for further research.

Table 11. Percentage of Responses in the Productive Strategies.

Strategy	Short Reprompt (% of responses)	Instructive Reprompt (% or responses)
Rephrased	6%	11%
Shorter	14%	18%
Fewer Ideas	8%	6%
Changed Request	4%	8%
TOTAL	32%	43%

7.   TURN-TAKING IN HUMAN-MACHINE DIALOGUES

In human-human dialogues turntaking is a complex negotiation between the participants with some known rules.   There are a variety of vocal mechanisms listeners and speakers will employ to indicate whether or not the speaker is finished and whether or not the listener is ready to take his or her turn.   Also, most obviously, the listener is interpreting the syntactic and semantic content of the speech from the speaker and can use this information to aid in the decision as to whether the speaker has completed his or her turn (Clark and Clark, 1977).

In contrast, in human-machine dialogues machines have relied mainly on pauses to determine if the speaker has finished.   That is, the system has some set timeout value (perhaps three seconds), through which the speaker must be silent, in order to signal the machine that it is its turn to speak.   This is an imperfect mechanism at best.   It is difficult to determine what value to use as the timeout value.   If this value is set too low (perhaps 1 second), then short pauses on the speaker's part are misinterpreted as the end of the turn and a muddled dialogue ensues.   For example:

System:	How may I help you?
Caller:	*Calling card call to 908 [pause 1 sec.] ...9*
System:	[system interrupts] What was that phone number?

This will be perceived as a very rude system.

The alternative is to set the timeout relatively high to avoid cutting in on a speaker's turn.   This also has some negative consequences.   The most obvious drawback is that the dialogue drags on at an unnaturally slow pace.   Callers get bored and frustrated that the system doesn't move along more quickly.   Also, some callers will attempt to speed up the transaction by providing more information per turn, which can reduce recognition accuracy and further slow the dialogue.   For example in a system where the timeout is set to three seconds a dialogue like the following might occur:

System:	How may I help you?
Caller:	*Calling card call. [Pause 2 seconds]Ah, to area code 908*
	949 1111. [Pause again 2 seconds]. C'mon you stupid
	machine.

Ultimately, the best solution to the turntaking problem with human-machine dialogues will be to better model the kinds of cues that humans use while in conversations with other humans, including the syntactic and semantic content of the utterance. The HMIHY system was not yet able to do this, so pause duration was used as the indicator that the speaker was finished with his or her turn.

With an early version of the HMIHY system there was some delay between the end of the caller's speech and the beginning of the next system prompt due to the length of time it took to process the caller's speech. The longer the system took processing the caller's speech, the better the analysis would be. On the other hand, if the machine takes too long to respond, the caller often gets worried that no one is on the line and begins to speak again (often with something like "Hello? Anyone there?"). The HMIHY system had the ability to cut off processing of the speech if it passed some period of time that was deemed too long to leave the caller waiting. The question then arose, what is too long? If the pause limit was set based upon the pauses in fluent human-human dialogue then the bar is set impossibly high, and this might not be the right comparison point. After all, the callers know that they are speaking to an automated system and not a human. It is reasonable that their expectations about the system behavior would be different from their expectations of a human operator.

In listening to examples in which the silence between turns had gone on too long for callers (and they resorted to saying "Hello?") it appeared that the callers' concern was that the system had somehow inadvertently disconnected them. That is, they seemed to need reassurance that the system was still there. One possibility to address this might be to play a prompt or some other kind of notification to let callers know that the system is still on the line, but is processing their request. Perhaps if the system offers this kind of information, callers might be even more tolerant of system delays. To determine the acceptable time callers will wait and to test this notion of "busy notification" a Wizard of Oz study was conducted and is described in the next section.

7.1   Caller Tolerance of System Delay

In order to determine what delay in turntaking was tolerable for users, several delays were tested and users were asked to rate each as acceptable or not. In addition, in order to determine whether the delays are more tolerable when the caller knows that the system is connected, but busy, a musical sound effect was developed that was played after the end of the caller's speech was detected, but before the next prompt was played. At the end of each call, callers were asked to judge whether or not the delay they experienced was acceptable or unacceptable.

The pause that followed the caller's speech was varied from two to eight seconds, and whether or not the pause was filled with the sound effect or whether it was silent was also varied. Each caller experienced each level of delay with both the sound effect and silence, resulting in 14 calls per participant. The trials were presented in a random order. The test included 45 participants.

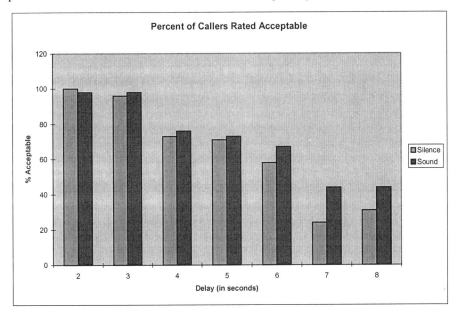

Figure 2. Percentage of Callers Who Judged the Delay Acceptable.

As shown in Figure 2, most callers found delays of two to three seconds to be acceptable. Furthermore, the sound effect made little difference for these short durations. It was only in the cases where the delay was unusually long, over six seconds, that the sound effect appeared to confer any real advantage in terms of improving the acceptability of the delay. In those cases, fewer than 50% of the participants found the delays acceptable even with the sound effect. Based upon these data, it was concluded that delays should be kept to less than three seconds and that a sound effect was not necessary or useful.

Much work remains to be done in the area of turntaking in natural dialogues between human and machine. Until more effort is made to imbue the machine with some of the grammatical and semantic knowledge that a human listener has, dialogues between humans and machines will have a stilted, slow quality.

8. SUMMARY

As speech recognition and natural language understanding technologies mature, spoken dialogue systems will begin to replace existing IVR systems and will also create new opportunities for automation. As the technology matures, designers of this new breed of automated systems will need to develop a new set of design guidelines that fit the capabilities of these new systems. This chapter has reported on some early research conducted at AT&T to begin to this area.

The initial finding from these studies was that making conversations more natural and anthropomorphic is good. Casual, anthropomorphic prompts can be shorter and more efficient. Users did not mind the casualness of the prompts that were tested. Anthropomorphism resulted in higher customer satisfaction. In general, our attempt to design the prompts to model the speech of human operators resulted in a system that was human sounding and well liked by callers.

Even with a fairly advanced spoken dialogue system there will be occasions in the dialogue that cannot be modeled after human-to-human conversation, so there will always be the need for some "non-natural" portions of the dialogue. The research summarized in this chapter indicated solutions for some of these "non-natural" dialogues. First, users will naturally produce better responses if it is impressed upon them that they are using an automated system. Second, reprompts that explicitly indicate that a different response is needed will produce better speech on the second try. Third, the best confirmation strategy might be one that explicitly requires a yes/no response.

Despite the need for some "non-natural" portions of the dialogue, natural spoken dialogue systems should provide users with greater flexibility and ease of use for most complicated tasks. As this technology becomes more advanced and moves out of the research lab into real applications, users should begin to experience the benefit of shifting more of the effort to the machine, making communication with an automated system easier.

REFERENCES

Brems, D. J., Rabin, M. D., and Waggett, J. L. (1995). Using natural language conventions in the user interface design of automatic speech recognition systems. *Human Factors, 37*(2), 265-282.

Clark, H. H., and Clark, E. V. (1977). *Psychology and language*. New York: Harcourt Brace Jovanovich.

Clark, H. H., and Schaefer, E. F. (1987). Collaborating on contributions to conversations. *Language and Cognitive Processes, 2* (1), 19-41.

Day, M. C., and Boyce, S. J. (1993). Human factors in human-computer system design. In M. Yovitz (Ed.), *Advances in Computers* (pp. 333-430). San Diego: Academic Press.

Falzon, P. (1990). Human-computer interaction: Lessons from human-human communication. In P. Falzon (Ed.), *Cognitive ergonomics: Understanding, learning and designing human-computer interaction* (pp. 51-66). London: Academic Press.

Franzke, M., and Marx, A. N. (1993). Is speech recognition usable? An exploration of the usability of a speech-based voice mail interface. *SIGCHI Bulletin, 25*, 49-51.

Gorin, A.L. (1995). On automated language acquisition. *Journal of the Acoustical Society of America, 97*(6), 3441-3461.

Gorin, A. L., Parker, B. A., Sachs, R. M., and Wilpon, J. G. (1996, October). "How may I help you?". *Proceedings of the IVTTA* (pp. 57-60). Piscataway, NJ: IEEE.

Gould, J. D., Conti, J., and Hovanyecz, T. (1983). Composing letters with a simulated listening typewriter. *Communications of the ACM, 26*, 295-308.

Kennedy, A., and Wilkes, A. (1988). Dialogue with machines. *Cognition, 30,* 37-72.

Richards, M. A., and Underwood K.(1984, April). Talking to machines: How are people naturally inclined to speak? *Contemporary Ergonomics 1984: Proceedings of the Ergonomics Society's Conference* 2-5 (pp. 62-67). London: Taylor & Francis.

Shneiderman, B. (1992). *Designing the user interface* (2nd ed.). Reading MA: Addison Wesley.

---

[1] The members of the research team included Allen Gorin, Egbert Ammicht, Alicia Abella, Tirso Alonso, Barry Parker, Jim Scherer, Guiseppe Riccardi, and Jerry Wright. The work reported here would not have been possible without the efforts of this remarkable team.

[2] A brief note about the length of this prompt: Designing the user interface for a system which is used by a particular user only occasionally is distinctly different than designing for a frequently-used, subscriber-based system, like a voice messaging system. In the former case, the prompts could be somewhat wordier and more informative, since the users are likely to be novices and in need of more assistance, and it is assumed that the users will not repeatedly use the system so as to find the length of the prompts annoying.

3 EVALUATING THE QUALITY OF SYNTHETIC SPEECH

Alexander L. Francis* and Howard C. Nusbaum

Department of Psychology
The University of Chicago

Key words: Synthetic Speech, Text-to-Speech (TTS), Intelligibility, Naturalness

Abstract: *The utility of text-to-speech systems depends on two aspects of their performance: intelligibility and nateralness. Evaluation of these aspects of synthetic speech provides important information about the performance of a speech synthesizer in comparison to competing products. It can be important for developers to understand where the realitve strengths and weaknesses of a particular synthesizer are so they can assist in development as well as marketing. Diagnostic evaluation directly can assist in the development effort by pinpointing specific problems in synthesis that can be redressed by engineering solutions. This chapter will outline the prinicles that govern the design of tests to measurethe performance of a text-to-speech system. Moreover, we will discuss the factors that influence performance on these tests. Finally, we will review some performance comparison tests used for text-to speech systems.*

1. INTRODUCTION

It has been nearly 20 years since Dennis Klatt's (1980) publication of "Software for a cascade/parallel formant synthesizer" paved the way for the development of relatively low-cost, relatively intelligible speech synthesis from parametric descriptions of speech. Starting with MITTalk (Allen, Hunnicutt, Carlson and Granstrom, 1979), the development of a system that would "read aloud" unrestricted text promised to revolutionize human-machine interaction. Developers imagined that one might soon be able to converse with every manner of mechanical device, from vending machines and cash registers to computers and flight instruments (Nusbaum and Pisoni, 1985). And why not? Speech is the natural mode of communication among humans, and it is not unreasonable to expect that information might be more efficiently transmitted from machines to human users via speech, rather than through lights, bells, whistles, meters, or text displays.

However, despite the widespread availability of numerous text-to-speech (TTS) systems (some still based on Klatt's original work, and others using considerably different synthesis methods) synthetic speech has not yet lived up to its initial promise. Computer-generated speech is still difficult to understand, and recognition of synthetic speech by humans is slower and requires more effort than recognition of natural speech (Nusbaum and Pisoni, 1985). Furthermore, even high-quality synthetic speech still sounds mechanical and unnatural (Nusbaum, Francis, and Henly, 1995).

It is true that in certain domains coded speech has become common (e.g., cash registers at some food stores, some telephone assistance). However, in these situations a relatively small number of different messages must be transmitted with great frequency, making the use of unrestricted TTS systems unnecessary. Furthermore, the most common type of speech used in these domains is catenation of whole words or phrases encoded from speech produced by a human talker and resynthesized on demand. In contrast, the application of large (or unlimited) vocabulary TTS synthesis has been, for the most part, restricted to situations in which the full range of natural language is used, or the message set is variable and unpredictable, such as in reading assistance for the blind and vocal prostheses for the speaking impaired (cf. Hunnicutt, 1995).

This lack of general application of TTS synthesis is likely due to a number of disparate factors, only some of which are related to the actual quality of the speech. However, there is no denying that even the best systems currently sound much worse than natural or coded speech. Synthesis by rule still sounds quite unnatural, and such speech is often noticeably more difficult to understand than human speech.

In order to evaluate whether a TTS system is likely to be useful in a given task or application, it would, of course, be most informative to employ it in that context and then record the reactions and performance of a large number of real users. However, in practice such full-scale field tests are not generally possible. In some cases, expected field conditions are difficult or prohibitively expensive to replicate, while in other cases the development process demands repeated cycles of testing and modification. Therefore, in order to make the most out of smaller scale controlled tests, it is important to focus on particular aspects of speech communication (especially those most affected by the methods of speech synthesis) and develop tests that accurately and efficiently assess a system's performance on these characteristics.

In this chapter we will first consider the three most common attributes of speech quality, how these attributes are typically measured, and how they are useful. These measures are equally applicable to testing the quality of all kinds of synthetic speech as well as natural speech that has been processed (e.g. compressed) or transmitted (electronically or otherwise). However, in this chapter we will concentrate entirely on examples from the domain of text-to-speech synthesis, and some of the principles and tests discussed later make sense only in that context. For a review of more general factors relevant to evaluating the intelligibility and acceptability of all kinds of speech, see Schmidt-Nielsen (1995). Subsequently, we will outline the principles that govern the design of tests to measure the performance of text-to-speech systems. Finally, we will discuss some strengths and weaknesses of a number of existing standardized tests, and we will provide some examples of the development of testing methods from our own research.

2. THREE MEASURES OF SPEECH QUALITY

Traditionally, developers of TTS systems have distinguished between three basic measures of performance: acceptability (or preference), intelligibility, and naturalness (Hecker and Williams, 1966). In the context of laboratory evaluations of the quality of speech produced by a TTS system, these measures are intended as an indication of how successful the system will be at producing speech that 1) is generally usable in some overall sense, 2) transmits intended messages clearly and effectively, and 3) sounds like natural human speech when used in its intended application.

More recently, a further distinction has been made between intelligibility and comprehension (Ralston, Pisoni, and Mullennix, 1995). Ralston et al. (1995) argue that traditional measures of intelligibility do not take into account the full range of cognitive mechanisms available to listeners attempting to understand connected speech (synthetic or natural). They argue that there is a difference between the task of recognizing particular sounds, words, or phrases (traditionally used to assess intelligibility) and the task of actually understanding a speaker's intended message (the task of most users of most speech synthesis systems). Because comprehension and token recognition are different kinds of tasks requiring the application of at least some different cognitive abilities, traditional measures of intelligibility do not necessarily reflect accurately the degree to which a synthesizer is capable of transmitting intended messages accurately. This is because, even if segmental intelligibility is low, context, prosody and task constraints can help a listener understand a message. For example, if a synthesizer has a poor segmental intelligibility score because certain vowels are all produced identically, it might be difficult to identify a word such as PUT produced in isolation because the word heard as /p?t/ might be interpreted as PIT, PET, PUTT, or PUT. However, in the context of a sentence (in which these vowels are all still indeterminate) such as *He p?t th? sh?v?l in the sh?d.* the meaning of the sentence, and even the identity of the individual words becomes much clearer. Our perspective is that although it is possible to separate out the issues of segmental intelligibility and comprehension, the need to do so depends on the specific goals of the evaluation process. In this chapter, we specifically discuss problems of assessing comprehensibility, but, in the interest of maintaining continuity with the main body of work on evaluating synthetic speech, we will consider comprehension testing as subsumed under the more traditional term of intelligibility. In doing so we will propose a considerable expansion of the set of tests used for evaluating intelligibility in order to take into account the wider range of cognitive mechanisms available to human listeners for understanding spoken language.

2.1   Understanding traditional measures of speech quality

Before discussing these measures individually, it is important to consider some of the factors related to their nature and use. Often measures of acceptability, intelligibility, and naturalness are used in setting performance benchmarks for development or for selection of a specific TTS system for a particular application. However, this can dangerously oversimplify the evaluation processes. As we will discuss later in this section, successfully tailoring a system to achieve high performance on one or another particular test may result in a misleadingly positive assessment of that system's overall performance if only one test is used for assessment. Rather than simply accepting these three measures as external, absolute scales for setting performance standards, it is important to understand what they actually measure, and how the different components of a TTS system interact with listeners and task demands to affect performance on these measures.

These Measures Are Interdependent. Although they are often treated as measuring distinct aspects of a synthesizer, these qualities are not completely independent of one another. For example, since acceptability is a global measure of overall quality of speech, it depends on all the more specific perceptual qualities of the speech. Thus, if intelligibility or naturalness is low, then acceptability is not likely to be very high. Similarly, in many cases speech with low naturalness is not as intelligible as more natural sounding speech, while speech that is low in intelligibility is almost invariably perceived as unnatural. This correlation of naturalness and intelligibility has generally been assumed to be the case for all speech. However, in our laboratory we have recently come across a situation in which a voice is rated as highly natural, but is less intelligible than should be expected from the naturalness rating alone. Many of the factors that underlie any one of these measures (e.g., naturalness) may also be significant for the other two measures as well. For example, if a particular synthesizer fails to realize the co-articulation of temporally adjacent speech sounds in the same way as humans do, then the synthesizer will sound less natural because coarticulation is a natural consequence of the physical, physiological, and psychological constraints on producing speech with a human vocal tract. However, this synthesizer will also be more difficult to understand because coarticulation may contribute to the efficient encoding of speech, and listeners may expect coarticulation as part of the acoustic-phonetic structure of speech (Liberman, Cooper, Shankweiler, and Studdert-Kennedy, 1967).

These Measures Are Complex.   Besides being interdependent, these three measures are all complex, in the sense that they simultaneously reflect diverse characteristics of the synthesizer, linguistic properties of the text being spoken, and the momentary and characteristic states and properties of the human listener. Because they reflect the interaction of many different aspects of the synthesis process, the underlying factors which affect acceptability, naturalness and intelligibility are not all objectively recoverable from the speech signal alone. Thus, although the same synthesizer will likely always produce the same speech given the same text, performing two different tests of intelligibility is likely to result in two different scores. This happens because different tests place more or less emphasis on the role of different aspects of synthesis (Greenspan, Syrdal, and Bennett, in press). For example, it is likely that some synthesizers will be more acceptable to some listeners in some domains, and less acceptable to others in other contexts. Since, in English, spoken digits are less easily confused than spoken letters, lower intelligibility speech might be more acceptable in an application reporting sports scores or bank balances than in an application for spelling names. Thus, while it is possible to develop tests that primarily measure either acceptability, intelligibility or naturalness, it is important to keep in mind which aspects of the entire process of communication (including the text, the synthesizer and the listener) is actually being targeted. In many cases, the use of multiple tests (even of any one of these measures) will provide more reliable and valid results than a single test.

These Measures Are Relative. Furthermore, because laboratory testing conditions can seldom reflect all relevant aspects of actual application conditions perfectly, test results cannot be taken as absolute measures. That is, just because speech produced by a particular TTS system is identified 92% correctly on a particular test of intelligibility does not mean that people using that synthesizer in its intended context will always recognize 92% of the words it produces. Nor does it mean that speech produced by the synthesizer will necessarily be recognized 92% of the time on some other test of intelligibility (Williams and Hecker, 1968; Greenspan, et al., 1998). For this reason, it is important to use test results as relative scores. That is, synthesizers should always be tested in comparison with other reference speech whenever possible. Ideally, these reference voices should include both natural human speech and synthetic speech whose quality is well understood because it has been tested on many different tests already. When comparing two synthesizers, it is often most informative to simply compare the rank ordering of each synthesizer relative to the other, and to that of a natural human voice presented in the same testing conditions.

2.2   Acceptability

Acceptability is the most global assessment of the three measures of speech quality. Acceptability is determined primarily by asking users, or experiment participants, to listen to samples of speech such as sentences or longer spoken discourse produced by a synthesizer. The participants then rate or judge the acceptability -- or their preference for each utterance. In this kind of task, the subjects judge the overall quality of the speech. The instructions might focus the subjects on how much a particular task could benefit from this particular speech, or how acceptable the speech would be for use in the task. Subjects can respond in a number of different ways. Often, subjects either rate the acceptability for each utterance right after they hear it, or listen to a set of utterances select the most acceptable one. A third method is to present the subjects with stimulus sets, each consisting of the same utterance produced by different TTS systems. Subjects in this kind of test are then asked to explicitly rank each voice that they heard in order of acceptability for that utterance (e.g., Huggins and Nickerson, 1985). If overt acceptability ratings (rather than relative rankings) are elicited, they are generally made on an interval scale (e.g., from 1 to 10) or a rank-ordered scale (e.g., Excellent, Good, Fair, Poor, Horrible).

The texts for acceptability rating tasks may be drawn either from general literary sources or, when testing applications designed for specific purposes, from particular areas of discourse in which the application will be used. For example, if a TTS system is being developed for use in a voice-mail system, the corpus of utterances should reflect the particular requirements of that domain, such as a representative sample of common names, telephone numbers, e-mail headers, messages, and related information. On the other hand, if the system is being designed for incorporation into a hands-free direction system for use in an automobile, the test corpus should consist of words, sentences and phrases likely to be used in giving directions.

In this way, acceptability ratings can provide a general score that indicates how likely a synthesizer is to succeed as a product in a particular application. Although the utility of measuring acceptability is obvious (unacceptable synthesizers will fare poorly in market competition), acceptability is also the factor least amenable to accurate diagnostic testing in the laboratory. Acceptability is too gross a measure of quality to diagnose specific problems with a particular TTS system. Indeed, acceptability often depends on a great many factors, including subjective phenomena such as listeners' preconceived notions of how the voice of a talker in a particular context should sound. For example, during one test in our laboratory involving five male voices in many different service-oriented contexts such as banking information, weather reports, and movie times, many subjects expressed surprise that we did not use any female voices, because they expected to hear women speaking in these contexts! Although it does not appear that listeners' expectations in this regard had any consistent effect on measured acceptability judgments, other preconceived hypotheses about the "right" kind of talker for a particular domain might. Because of this subjective quality, it may be difficult to compare acceptability ratings across tests or across contexts. Because acceptability can be so context-specific, for laboratory testing it is often more practical to reduce the emphasis on overall acceptability testing in favor of focusing on the individual contributions of more basic system characteristics.

2.3   Intelligibility

Intelligibility is a more specific measure of how well particular linguistic information (sentences, words or phonemes) is transmitted by a TTS system. Again, the benefit of assessing intelligibility is clear. Unintelligible speech is useless, and paradoxically, low intelligibility speech can be worse. If listeners have to work hard to understand speech, they may become unduly fatigued, or they may choose to ignore the speech entirely. If the speech is easy to misunderstand (as opposed to merely hard to understand) listeners may make more incorrect decisions or incorrect responses in an application. On the other hand, if the speech is difficult to understand at all, listeners may focus so much attention on speech comprehension that they neglect other aspects of their task. In either case, their reactions to the speech will be slowed or inappropriate for the intended message, which can have serious consequences in many contexts.

However, while acceptability can be thought of as a more or less unitary, general score incorporating or reflecting all aspects of the performance of a synthesizer to some degree, intelligibility, while much more specific, is also much more complex to measure. Synthesizers differ in terms of: 1) the synthesis methods they use to produce speech sounds (e.g., cascade/parallel synthesis, such as is used by DECTalk, vs. diphone concatenation, such as is used by Apple PlainTalk for some of its voices); 2) the grapheme-to-phoneme and phoneme-to-sound rules that they use to convert text to speech; 3) the parsing algorithms used to derive intonation contours; and 4) the methods for imposing those contours on the segmental phonetics of the utterance. They may also differ in terms of the "sex" and "age" of the talker, and the size of the dictionary of names, foreign words, and other text strings with irregular pronunciations.

Consequently, synthesizers also differ in the degree to which these methods that they employ succeed in producing intelligible speech. Just because a particular synthesizer might have an excellent algorithm for producing intelligible segments (or diphones) does not mean that that system also necessarily will be good at producing appropriate cues to lexical stress in combination with those segments. Such a system might be able to produce more or less intelligible monosyllabic words, yet be considerably worse at producing most polysyllabic words intelligibly. The results of different tests may depend more or less heavily on one or another of these factors. For example, an intelligibility score derived from administering a test that consists only of transcribing monosyllabic words will reflect very little about how successful a system is at realizing sentential prosody, but it may be strongly influenced by the choice of methods for producing the individual segments of words in isolation. In contrast, a test of intelligibility relying on the recognition of polysyllabic target words might be much more influenced by aspects of synthesis related to prosody.

Furthermore, the results of such a test also are likely to be influenced by listeners' expectations based on their real-world knowledge and their knowledge of sentence structure and word and phrase frequency. For example, it is well known that more frequent words (words which occur frequently in spoken and written language) are recognized more quickly than less frequent words, and this should be taken into account when analyzing the results of quality testing involving word recognition or transcription. Other linguistic and extra-linguistic factors that influence intelligibility ratings will be discussed in more detail later in this chapter.

Although acceptability measures may be influenced by intelligibility, it is less likely that intelligibility will be influenced by acceptability. Intelligibility tests do not rely on the same kind of subjective assessment of speech quality used in acceptability tests. Instead, intelligibility tests measure recognition *performance*. Thus, listeners have to *identify* consonants, vowels, words, or sentences. While there are many different forms of intelligibility tests, these tests differ from acceptability tests in that speech is used to carry out a task, and performance on the task is measured. In contrast, in acceptability tests, listeners make a *judgment* about the quality of the speech. Thus, intelligibility tests may be more reflective of the way tasks are performed in an application based on the speech quality, while acceptability tests may measure listeners' preference for using the speech.

2.4   Naturalness

The third measure, naturalness, is a measure of how much a synthesizer sounds like human speech.   Naturalness has only recently become a serious concern for developing better speech synthesis.   As long as even the best synthesizers were relatively unintelligible, TTS developers had to be concerned primarily with improving intelligibility.   After all, speech is totally useless if it cannot be understood.   For the most part, there was little expectation of using TTS synthesis in more than a very limited set of contexts in which it scarcely mattered whether the synthesizer sounded human or not.   For example, if the only way a paralyzed person can communicate over the telephone is by controlling a TTS system with a joystick, it matters less that the voice sounds mechanical and artificial than that the TTS system is at least comprehensible to the listener and controllable by the user.   However, as intelligibility has improved, developers have begun to take the time to address this more aesthetic aspect of TTS synthesis.

Although the parameters that affect naturalness are much less well understood than those of intelligibility, there are also obvious benefits to producing more human sounding speech.   For example, for people with motor control disabilities or particular speech or language disorders, a synthetic voice may be the only method of spoken communication.   While intelligibility is certainly a concern for these users, the quality of the voice itself can also be significant.   For example, users may prefer to use a voice that "sounds like me."   Communicating with a clearly mechanical voice or a voice with the wrong gender or age may feel embarrassing or awkward for many users, and it may induce prejudicial or inappropriate assumptions in unprepared listeners. For example, Mirenda and Beukelman (1987) and Greene and Pisoni (1988) both found significant age by system interactions, suggesting that child listeners prefer to listen to TTS systems that sound like children (see also Logan, Greene, and Pisoni, 1989). In most applications, the more natural and human-like a voice sounds, the more likely it is to be accepted by users and listeners alike.

Interestingly, in some applications, especially those involving privacy or security issues, users may be more comfortable listening to a voice that is obviously unnatural, rather than one that sounds like a human talker. In a study performed in our lab, a small group of listeners rated a particular TTS system as the most preferable one (out of five) in a banking-by-phone context, although they considered that synthesizer to be less acceptable than the other four in many other contexts. They also tended to rate the synthesizer as sounding less human than the other synthesizers overall.   When asked why they rated the voice higher in this condition, these listeners generally reported that they would feel more comfortable knowing that their account balance had been accessed by a computer, not by a human being.   However, since it is quite clear that we know how to generate an *unnatural* sounding synthetic voice, there should be no problem implementing one in these applications, should the need arise.

3. PRIMARY CONSIDERATIONS IN MEASURING SPEECH QUALITY

For most purposes, it may be enough to simply evaluate acceptability, intelligibility, or naturalness as a rough estimate of the quality of a TTS system. However, for the purposes of improving existing systems and of developing new ones, it is useful to be able to determine why a system is not as acceptable, intelligible, or natural as desired. As we have discussed already, it is often the case that the same factors can affect both the intelligibility and naturalness (and thus almost by definition the acceptability) of a particular system. Many of the factors that contribute to the design of a TTS system may appear to bear little relation to the quality of the synthetic speech itself, but many design decisions made for reasons unrelated to speech quality can have an effect either directly on the quality of the speech or at least on the appropriate methods for evaluating speech quality. In this section we will briefly consider the components of a system of communication involving TTS synthesis (including the synthesizer, the text, and the listener) and we will discuss how these components affect the evaluation of the acceptability, intelligibility and naturalness of synthetic speech produced by rule.

3.1   The Synthesizer

When considering the evaluation a TTS system, there are two areas in which the system itself could be deficient. The first is in the actual signal characteristics of the voice. If the fine structure of the acoustic signal of synthetic speech does not closely match the source characteristics of natural speech, listeners can detect this. Similarly, if these source characteristics do not vary in a natural manner, listeners may find the speech unacceptable or unnatural (Nusbaum et al., 1995).

The second area in which a TTS synthesizer could be deficient is in terms of the linguistic rules used to convert text to speech, including rules for converting text to a phonological representation (spelling-to-sound rules and rules for computing prosody) and rules for converting the phonological representation of an utterance into acoustic patterns (phonetic implementation rules). Failure to correctly implement these aspects of the human voice in a TTS system contributes to the unnatural quality of synthetic speech (Nusbaum et al., 1995), and may detract from its intelligibility. (For spelling-to-sound rules, see Fant (1991) and Nusbaum and Pisoni (1985). For prosodic factors, see Klatt (1976) and Syrdal (1989).)

Voice Quality. In natural human speech, there are many parameters that contribute to the perception of voice quality but are not known to be useful for making linguistic distinctions in English.   These parameters include glottal source characteristics such as the ratio of aspiration noise to periodic sound in higher harmonics and the relative amplitude of the fundamental frequency (f_0), and filter characteristics such as the bandwidths of lower formants and the location of poles and zeros in the vocal tract transfer function (Klatt and Klatt, 1990). Differences in these characteristics are easily perceived by listeners as, among other things, differences between male, female, and child voices or as differences between voice qualities such as creakiness or breathiness (Klatt and Klatt, 1990; Laver, 1980).

Despite increasing sophistication in identifying acoustic factors that contribute to voice quality and reproducing them in synthetic speech (e.g., Klatt and Klatt, 1990), there are many characteristics of the human voice that are still inadequately modeled in the speech of TTS synthesizers. For example, the shape of the glottal waveform tends to be more similar from period to period in synthetic speech than in natural speech. In the latter, the waveforms of the glottal pulses are considerably more variable, and listeners appear to be sensitive to this natural pattern of variability (Fant, 1991; Klatt and Klatt, 1990). Similarly, prosodic characteristics such as pitch, amplitude and segmental and syllabic durations vary not just in linguistically meaningful ways, but also to reflect emotional and other affective characteristics of speech. Clearly, even the best synthetic speech is not yet able to reflect the wide range of affective expression available to human talkers. However, listeners' judgments of speech quality may be substantially influenced by the perceived naturalness of synthetic speech, even if the listening task merely involves recognition (Nusbaum et al., 1995).

Linguistic Cues. The first step in converting text to speech is to determine the appropriate pronunciations of text. This process of grapheme-to-phoneme conversion can have a significant effect on the acceptability of a TTS system. For example, the grapheme string *St.* can be pronounced as either *Saint*, as in *The Bells of St. Clements*, or as *Street*, as in *Anderson St.* A synthesizer that is not able to produce the correct pronunciation of a text string in the right context will be more difficult to use. In order to evaluate the degree to which the language model is capable of correctly analyzing the grammatical and semantic context in which such ambiguous strings occur, a complete test corpus should include a representative sample of such strings.

Once the language model and grapheme-to-phoneme (or grapheme-to-diphone) rules have been applied to the text to derive a phonological representation of the speech that should be produced, the next task is to map this phonological information onto acoustic patterns. Because the mapping from phonology to acoustic patterns is so complex, the methods employed at this stage can have a profound effect on the intelligibility, naturalness, and acceptability of synthetic speech, and therefore these factors tend to be the ones most commonly studied or evaluated (Nusbaum and Pisoni, 1985).

Natural speech is extremely rich in terms of the patterns of acoustic phenomena that can be used by talkers to produce exemplars of linguistic categories. For example, numerous correlates of a stop consonant such as /b/ or /d/ have been identified in spoken American English (cf. Olive, Greenwood, and Coleman, 1993). These include: 1) correlates of the manner of articulation, including the presence of a burst of transient noise in certain contexts and the abrupt rise of amplitude with the onset of voicing; 2) correlates related to the place of articulation, such as the spectrum of the transient noise of the burst-release (Blumstein and Stevens, 1981) and the slope and starting frequency of the formant transitions into the vowel (Cooper, Delattre, Liberman, Borst, and Gerstman, 1952; Dorman, Studdert-Kennedy, and Raphael, 1977); and 3) correlates of voicedness, such as the duration of the silence between the burst-release and the onset of phonation (Lisker, 1978), the onset frequency of the first formant (Lisker, 1975), and the duration of the preceding vowel (Port and Dalby, 1982). Much of the research on acoustic cue structure for the purposes of developing synthetic speech has concentrated on identifying acoustic patterns in human speech, such as those described by Lisker (1978), that correlate with the *production* of particular linguistic categories (such as phonemes). In contrast, most psychological work on speech perception has concentrated on discovering how particular acoustic patterns cue the *perception* of particular categories. It is important to distinguish between acoustic *correlates* of linguistic categories, which are sound patterns that are predictably observed in the production of a particular category in a particular context, and acoustic *cues*, which are correlates that, when present in the speech signal, induce the *perception* of a particular category in a particular context.

For example, Lisker (1978) describes 16 different acoustic correlates that distinguish /b/ from /p/. However, although most of these correlates are likely to be present in any given production of a /b/, they need not all appear for the sound to still be perceived as a /b/ as opposed to a /p/. In fact, it has been demonstrated that certain correlates such as the second formant transitions into the vowel, can function alone to cue the place of articulation, another feature that distinguishes different linguistic categories such as /b/ vs. /d/ vs. /g/ (Delattre, Liberman and Cooper, 1955).

Furthermore, it may be the case that any correlate can conceivably function as a cue in the right circumstances. Thus, if other, more commonly used cues are missing or misleading, listeners might be able to use their knowledge that some other acoustic pattern covaried with the more familiar cues to recognize linguistic categories on the basis of the presence of these covarying cues. For example, in American English, vowels tend to be longer when followed by a voiced consonant (as in the words *his* or *rabid*) than they are in an otherwise identical context preceding a voiceless consonant (as in the words *hiss* or *rapid*) (Denes, 1955; Port and Dalby, 1982). In natural human speech the voicing characteristics of the following consonant can be uniquely determined independently of the vowel duration by means of other cues such as the presence or absence of a voicing fundamental (f_0). If a TTS system fails to supply the appropriate cues to the identity of the consonant, but succeeds in replicating the natural covariance of other acoustic-phonetic cues such as the duration of the preceding vowel, then listeners will likely be able to make the necessary distinctions. However, if the synthesizer fails to produce both the consonantal cues and the vowel-length cues to consonant identity, the resulting speech will sound odd, and naturalness and intelligibility may suffer. Thus, an obvious strategy in developing synthetic speech might be to attempt to replicate, as completely as possible, all of the correlates that are observed in natural speech.

Unfortunately, the role of particular correlates in cueing the perception of phonetic categories is not static. In some cases, decreasing the influence of one correlate can increase listeners' reliance on another correlate (Fitch, Halwes, Erickson, and Liberman, 1980; Repp, 1982). Repp (1982) give a number of different examples in which different acoustic correlates of various segmental features such as voicing and place of articulation can be "traded" in synthesized speech sounds in order to maintain the perception of the same phoneme. For example, starting with an acoustic signal that is phonetically ambiguous between a voiced and a voiceless stop (i.e., the signal is identified equally often as a voiced stop and as a voiceless stop), it is possible to increase the duration of the silent period between the burst release and the onset of phonation (the voice onset time, or VOT) such that the stimulus then is heard unambiguously as a voiceless stop. If, however, one also decreases the onset frequency of the first formant of the subsequent vowel to an appropriate degree, then the token will continue to be identified equally often as voiced or voiceless. In other words, Repp (1982) argues, these two cues are perceptually integrable, and the modification of one may be offset by a concomitant modification in the other.

However, despite the best efforts of speech synthesis researchers, even the most intelligible synthetic speech is still impoverished in terms of the number of cues and in terms of the complexity of systematic covariance of these cues as compared to natural speech. It seems likely that synthetic speech may never be able to replicate the full richness of cue structure and context-dependent covariance of natural speech. However, this does not mean that synthetic speech must necessarily remain less intelligible than natural speech. By incorporating an understanding of human speech *perception*, developers can decide which cues can be left out in which contexts without reducing intelligibility.

For example, some cues are more salient than others. If two cues - such as the burst-release spectrum and formant transitions of a synthesized stop consonant - are in conflict, listeners tend to base their judgments about the identity of the consonant on the information provided by the formant transitions, rather than the burst release cue (Walley and Carrell, 1983). This does not mean that the frequency spectrum of the burst release cannot function as a cue -- research in our lab has demonstrated that listeners can be trained to base their judgments on the burst release cue and ignore the formant transition cues (Baldwin, Francis and Nusbaum, submitted). However, even in this case listeners improve *more* with the same kind of training to listen to formant transition cues. That is, although listeners are able to adapt to using less salient cues, it takes more effort, and may still be less successful.

Just as segmental cues can interact in complex trading relations, prosodic parameters such as contrastive stress, and question and list intonation (to name just a few) also interact in significant ways with the realization of segmental cues. For example, in English vowels in contrastively stressed words tend to have a greater duration, pitch and amplitude than the same vowel in the same word in an unstressed context (Pickett, 1980). Thus, in natural speech, there may be a durational difference between a given vowel /ɛ/ in an unstressed position, such as in the word *Bev* in the statement *Bev loves Bob, not Bill*, as opposed to in a stressed position such as in the statement *Not Sally! Bev loves Bob*. Furthermore, this durational difference due to prosodic stress is independent from, but interacts with, the durational difference due to the voicing characteristics of the subsequent consonant (as discussed previously). Again, failure to replicate the full range of context-dependent variability in cue structure, including the effects of sentential prosodic characteristics, can make the speech of a TTS synthesizer sound less natural, less intelligible and less acceptable. A thorough evaluation of speech quality should include test stimuli having a wide variety of sentence structure and utterance type, including questions and statements with and without contrastive stress.

3.2   The Text

The ability of the synthesizer to produce the expected acoustic-phonetic cues to the linguistic content of a text is only one of the components of a TTS application that must be considered in evaluating speech quality. The nature of the message to be produced by the synthesizer, and subjects' knowledge of it, can also have a significant effect on the results of testing. The most important factors to consider are the size and composition of the message set and the amount of linguistic context provided to the listener by means other than the targeted speech itself.

Message Set. The nature of the text used in a listening test can have a profound effect on the results of the test. For this reason, in the case of acceptability testing it is usually preferable to match aspects of the listening task to the expected use of the final TTS application. For example, some synthesizers take only alphanumeric text as input, while others are specifically designed to deal with the characters and abbreviations in special domains such as reading electronic mail headers. It would not be sufficient to test an e-mail-reading synthesizer using only alphanumeric text, because that would not provide any information about the quality of one of its more important capabilities. In other words, if the application will be expected to read e-mail, the acceptability test should use materials structured like e-mail, and test subjects should be prepared to judge the speech in the context of listening to e-mail.

In contrast, for diagnostic testing during development, it may be more useful to select specific kinds of message sets in order to test specific parameters of the synthesizer more thoroughly. Thus, while some intelligibility tests require listeners to transcribe words heard in isolation or in predictable carrier sentences in order to diagnose segmental intelligibility, others involve transcribing or responding to whole sentences in order to tests syntactic and prosodic aspects of synthesis. Using a combination of these two kinds of tests will provide a more thorough evaluation of the intelligibility of a TTS system.

However, in addition to matching the messages used in testing to those used in application, it can be important to control other aspects of the message set as well. For example, it is well known that words which appear more frequently in spoken and written language are more easily recognized than infrequent words under similar conditions (Forster, 1976; Morton, 1969). It has been proposed that such frequency effects are in part due to the greater familiarity (experienced frequency) of common words (Pisoni, Nusbaum, Luce, and Slowiaczek, 1984). Therefore, listeners familiar with the words that make up a message are more likely to find that message intelligible than are listeners unfamiliar with those same words. Thus, when conducting an evaluation of the relative applicability of various TTS systems for use by experts in a field rich in jargon (for example, computer programmers or physicians), using test subjects who have little experience with the key words from that field may result in underestimation of the intelligibility of the synthesizer for its intended users. On the other hand, when performing diagnostic tests in the course of development, it is important to control the familiarity of words used in testing, so that intelligibility scores are not overestimated for message sets containing a high proportion of familiar words, or underestimated for message sets containing a high proportion of unfamiliar words.

The frequency of words within a particular test or trial on a test can also have an effect on measures of speech quality. For example, a small (closed) set of familiar and/or expected tokens will be recognized, in general, better than an open/unlimited set. Thus, if in a particular trial of an intelligibility test a synthesizer produces the word *bed* and listeners are asked to choose between the possible words *bed*, *red*, *fed*, *wed*, *led*, and *shed*, then there is immediately a one in six chance of the listener picking the correct answer, even if the word is totally unintelligible. In contrast, with an open response procedure, listeners can write down whatever they think they heard, choosing from among the set of all possible words they know. Besides reducing the chance of a correct guess, open response set tests can also provide information about which of the words' phonemes are not being produced correctly. For example, in the case described here, if listeners consistently replied that they heard the word "fed" rather than "bed" one might conclude that the synthesizer is failing to produce the segment /b/ appropriately (perhaps with too much frication). While such diagnostic information can be a useful result of using open response set tests, as we will discuss in the section on intelligibility testing, similar information can also be obtained by using the various Rhyme Tests. The latter usually have a closed-response format. Phoneme-specific sentence tests can also be used and are most easily conducted as open response tests.

In certain applications, word frequency can also have an effect on *production* of speech by a TTS system. For example, in the case of augmentative or assistive speech systems the mechanism for determining the text to be produced may be designed to be used easily by people with impaired motor abilities. Many such systems make use of a joystick-like controller coupled to a frequency-ranked lexicon and model of English syntax. Moving the controller allows words (and even phrases) to be selected for production from a display list made up of the words most likely to appear in the context of those words which already have been selected or produced. It is advisable to test such systems as a whole, rather than to evaluate the voice alone, because the production method may affect speech quality in significant ways. For example, it can disrupt discourse prosody at unexpected intervals as the user searches the lexicon for an infrequent word or a word that has been inappropriately left out of the selection list by an inadequate language model.

Linguistic Context. Even though the cue structure of synthetic speech can be impoverished or misleading, listeners are able to compensate by inferring missing information or reinterpreting incorrect information using other aspects of the speech signal, combined with their knowledge of the language. For example, recognizing words produced in sensible, contextually well-situated sentences is much easier than recognizing isolated words because most sentences provide many more cues to word identity beyond the segmental and prosodic characteristics of the target word (Miller and Isard, 1963; Nusbaum and Pisoni, 1985). For example, given the sentence "She likes her coffee with cream and ____." the most likely target word is clearly *sugar*. Of course, *sugar* is not the only possible word that could fit in this context, but it is the most likely one. Furthermore, not all sentences constrain the identity of their constituent words so severely as this. In the section on sentence testing later in this chapter we discuss the Harvard sentences, which were specifically designed such that one cannot easily predict some of the words in them given knowledge of the others. However, presenting whole sentences to listeners may reduce naturalness or acceptability ratings if sentential prosody is poorly realized, as discussed in the previous section. Intelligibility may also suffer, especially if sentential prosody is so poor that listeners are misled about the likely identity of the target word. A compromise of sorts may be found in the use of carrier sentences, such as "Please write _____ here," which provide a constant sentential intonation but do not provide semantic or pragmatic cues with which listeners can apply their real-world knowledge and linguistic experience to decipher otherwise unintelligible words.

Similarly, given a particular constellation of acoustic correlates that is not uniquely interpretable, listeners can use the context of the other acoustic patterns with which those correlates appear as a clue to what the talker likely intended to say. For example, Ganong (1980) presented listeners with an acoustic pattern that is midway between a pattern normally heard unambiguously as /d/ and one normally heard unambiguously as /t/ in two different contexts. When this ambiguous sound was presented in the first context, preceding the syllable *ask*, listeners were much more likely to report that they heard the word *task* than the acoustically equally plausible non-word *dask*. On the other hand, when exactly the same initial acoustic cue was presented preceding the syllable *ash*, listeners were much more likely to report having heard the word *dash* rather than the non-word *tash*. In both cases listeners used their knowledge of possible words in English to disambiguate the ambiguous cue pattern such that it was heard as if it were an unambiguous representative of the sound which would make the stimulus into a word.

3.3   Listening conditions

There are a number of factors that can influence evaluations of the quality of synthetic speech that have nothing to do with the actual generation of that speech. In particular, it is important to consider the conditions under which the speech is heard and the demands of the task(s) being performed while listening to the speech.

Acoustic Environment. Although synthetic speech is usually more difficult to understand than natural human speech, many of the environments in which it might be employed are not optimal, even for listening to human speech. Factory floors, automobiles, and aircraft cockpits may be too noisy for effortless listening, and various telephone systems may be so limited in frequency bandwidth that some acoustic cues are simply not transmitted.

In many cases of listening to synthetic speech in a noisy environment, listeners may have the option of simply increasing the volume of the synthesizer in order to make the speech more salient. However, if the synthesizer is to be used in a particular environment with known noise or frequency bandwidth characteristics (such as a factory floor with a high level of predictable background noise, or over the telephone), a preferable alternative may be to modify aspects of the voice to be maximally intelligible under those particular circumstances. For example, when human talkers speak in a high noise background, they certainly speak louder, but they also raise their pitch and increase the spectral tilt of their speech. In either case, testing the synthesizer in a noise-free environment at normal listening levels is not likely to provide a useful predictor of the quality of that synthesizer in that environment.

If the particular characteristics of the background noise spectrum are known, it is possible to include these measures in the testing environment to insure that the voice is still intelligible or acceptable under those conditions. For example, if the TTS system is intended for use over the phone, speech can tested through the appropriately matched band-limited filter. Similarly, if the TTS system is intended for use in a noisy environment, during testing the speech can be masked with noise matching the environmental noise in frequency and amplitude characteristics. In the case of acceptability testing, however, when including noise from the application environment in the testing environment it is important to make it clear to test participants that the noise is not a product of the synthesizer itself, but rather a constant characteristic of the environment in which it is to be used. Otherwise listeners may decide that the synthesizer is less acceptable *because* of the presence of noise. Testing using multiple reference voices (natural and synthetic) as we have discussed in the section on intelligibility, is another way of reducing the effect of such judgments.

Task Demands. Similarly, just as the intended use of a TTS system may vary, the complexity of the tasks carried out using the synthesizer may vary as well, and this should also be reflected in testing. For example, a system designed to be used for providing feedback to visually impaired computer programmers might make quite different demands on listeners than a system designed to be used by sighted, but very busy air-traffic controllers. In the first case, the task to be accomplished by listening is very complex. However, the user is not likely to be attempting to carry out any other tasks simultaneously. In contrast, TTS applications for air traffic control are more likely to serve an augmentative role, providing information in a audible modality that there is not time or attention available to receive visually. Thus, in the second case, although the listening task itself might be very simple, the fact that the listener is likely to be trying to accomplish a large number of other tasks simultaneously must also be taken into account when evaluating the quality of the speech used in the listening task. This problem is directly related to the role of limitations on users, especially with respect to working memory and attention, as discussed in the next section.

3.4   The User

Human listeners come with numerous expectations and preferences about how speech is supposed to sound, and what words and phrases talkers will use under particular conditions. They may also have certain social expectations about how they will be spoken to, and who will be speaking to them, in particular domains. Such pragmatic expectations may play a role in the comprehension of spoken language and, therefore, may also play a significant role in evaluating the quality of synthetic speech. For example, listeners inexperienced with synthetic speech may expect machines to sound like HAL from *2001*, or like C3PO from *Star Wars*. When synthetic speech fails to live up to these expectations, it is likely to be considered less acceptable.

Similarly, at the segmental level listeners may have developed implicit assumptions about how acoustic cues tend to pattern together in natural speech as a result of many years' experience listening to natural speech, as discussed in the section on acoustic-phonetic cues. Violating any of these expectations, either by failing to include sufficient cues or by producing misleading cues, may make more demands on listeners' attention, and may distract listeners from the performance of their tasks.

Expectations. Users of a TTS system may have particular expectations about how computers should sound. Similarly, they may have particular expectations about how interlocutors in particular situations should sound, and they may not expect, or want, to hear a computer talking to them in particular situations. It is well known that people often hang up without leaving a message when reaching an answering machine or voice-mail system, and it is likely that any general dislike of interacting with machines extends to interacting with TTS applications as well. However, even among users who are not *a priori* opposed to interacting with machines, in some interactive situations listening to a machine may be considered more acceptable than in others. For example, when requesting information or assistance, such as when asking for driving directions or computer support, users may prefer to listen to a human-sounding voice, under the assumption that more mechanical sounding voices are less likely to "know" how to solve the problem at hand. In contrast, as mentioned previously, when listening to sensitive personal information such as a bank balance, users may prefer to hear a less natural voice because they want to be sure that the information remains private.

Secondly, listeners may also have expectations about how particular text strings are pronounced depending on the application context. For example, a system that is designed to pronounce financial information in a maximally intelligible way may be relatively poor at producing telephone numbers correctly. In the first context, the number 367 should probably be produced as *three hundred and sixty seven*, whereas for a phone number it should be "spelled out" as *three six seven*. The outcome of highly context-dependent tests such as acceptability tests may reflect these expectations, such that, for example, in one context the TTS system that produces *three hundred and sixty seven* will be rated as more acceptable than one which produces *three six seven*. However, the relative rankings will be reversed in a different context. Thus, when measuring acceptability scores for a general purpose TTS system, it is advisable to request ratings from listeners in a wide variety of contexts so that no single factor has an overwhelming effect on the score.

Experience. Just as the domain of application has an effect on how listeners expect a system to speak, listeners' experience with language and with task domains affects the way listeners attend to speech. For example, every field of expertise has its own jargon and idiosyncratic phrases that may not make sense to listeners from outside the field who are not expecting those particular phrases. Thus, when testing a system designed to provide information within a specific field with specialized terminology such as investment banking or networking technology, it is be important to use test subjects who are familiar with the technical terms, abbreviations and phrasing peculiar to that domain. If the users of a TTS system are likely to be familiar with the particular topics to which the synthetic speech pertains, then it is important to insure that test subjects have access to the same knowledge.

On the other hand, users of a TTS system will not have access to the same particular information as test subjects, and test results may be affected by a particular pool of subjects' personal experiences. For example, in designing a test of a system for providing driving directions, it would be important *not* to use sample directions from a region or city that is familiar to the test subjects, because real users are less likely to ask for directions in a familiar area. If the test subjects already know how to get to the places the sample directions are describing, they may find the speech to be more acceptable or intelligible than if they really had to rely only on the information in the signal. In this case, a better test would be to use sample directions from an area which is known to be completely unfamiliar to the test subjects.

Attention And Learning. Experience with a particular talker also affects listeners' performance on intelligibility tests. As little as four hours of training with synthetic speech results in a significantly increased rate of recognition (Greenspan, Nusbaum, and Pisoni, 1988; Schwab, Nusbaum, and Pisoni, 1985; Lee and Nusbaum, 1989), and some improvement is noticable after considerably less training. Some testing methods take this into account, and disregard the first few trials in each condition (Gardner-Bonneau, personal communication, 1998). Many standard tests of intelligibility require that listeners be trained on all aspects of the evaluation task, including familiarity with all of the words used in testing, and with all aspects of the response task, such as pressing buttons or typing in words (ANSI, 1989). In training listeners for the purposes of testing the quality of TTS synthesis, it is important to know whether or not the future users of the application will be expected to be familiar with the synthetic speech. If the application is being developed to be used by naive users with little or no experience with synthetic speech, then it is important not only that the listeners in the evaluation experiments not be familiar with synthetic speech, but that all task-related training prior to testing be carried out using natural human speech, so that the test listeners do not develop expertise in understanding the voice they will be evaluating.

Another factor that must be considered when discussing training listeners is that listening to synthetic speech takes more attention than does listening to natural speech (Luce, Feustel, and Pisoni, 1983; Nusbaum and Pisoni, 1985). For example, Luce et al. (1983) found that listeners were significantly worse at remembering lists of synthesized words than they were at remembering lists of words produced by human talkers. Furthermore, they found that listeners' performance on other tasks such as remembering lists of digits decreased more when those tasks were carried out in conjunction with remembering synthesized words as opposed to with naturally spoken words. Finally, they also found that, when recalling longer lists of words, subjects' performance at remembering words at the beginning of the list (the primacy portion) was significantly worse for synthesized words than it was for natural speech. Luce et al. (1983) argue that these results suggest that processing synthetic speech incurs a greater demand on working memory than does listening to natural speech. In many cases in which synthetic speech could be employed, causing undue distraction or memory load is undesirable because users may have enough to keep track of without also having to devote attention to understanding messages (e.g., aircraft pilots, air traffic controllers, automobile drivers in traffic).

However, this does not necessarily mean that synthetic speech cannot be used in these environments. As mentioned before, listeners can learn to improve their recognition rates of synthetic speech with comparatively little training. As we have discovered in our lab, such training also reduces the demands on working memory or attention. In a study by Lee and Nusbaum (1989) listeners were tested on a word-recall task using synthesized stimuli under low and high levels of cognitive load before and after training on the TTS system that was used to produce the test stimuli. In the pre- and post-test evaluations of the attentional demands of recognizing synthetic speech listeners performed two tasks, a listening task and a secondary cognitive load manipulation task. Cognitive load was manipulated by use of a digit pre-load task based on the one described by Baddeley and Hitch (1974). In the low load condition, listeners were shown two two-digit numbers to remember during the listening task, while in the high-load condition listeners saw four two-digit numbers to remember. In each load condition the listening task was the same -- a speeded monitoring task in which listeners monitored for a target stimulus in a list of distractors (Logan, 1979). The effect of secondary memory load on the task of listening to synthetic speech was measured in terms of the difference in percent of targets recognized correctly in the low load condition vs. in the high load condition. As predicted by the hypothesis that listening to synthetic speech requires the commitment of working memory, listeners performed worse on the listening task when they had less memory available (in the high secondary load condition) than they did when they had more memory available (in the low secondary load condition). However, after training listeners were significantly faster on the word monitoring task in the low load, but not in the high load, condition. Lee and Nusbaum (1989) argue that this pattern of results suggests that, through training, listeners learn to direct available attention better. Thus, in the low load condition after training available attention is directed more efficiently than before training, resulting in the observed decrease in response time. However, in the high load condition there was less attentional capacity available overll and, therefore, the improvement in attentional distribution had no observable effect on response time.

Training on synthetic speech can decrease the amount of attention required for recognition, as long as there is sufficient attention still available. Thus, using training as a technique for compensating for the increased attentional demands of listening to the impoverished cue structure of synthetic speech should be considered only in cases in which the secondary tasks that must be performed in conjunction with the primary listening tasks do not require a great deal of attention. Furthermore, subjects used in an assessment task should be matched, as much as possible, with the expected users of the final TTS application in terms of their degree of experience with synthetic speech and with the secondary tasks associated with the listening task.

Finally, the role of attention must also be considered when designing tests. We have already discussed the fact that sentential and phonological context and the size of the set of response choices can all serve to facilitate recognition by limiting the set of possible responses. However, these contexts also provide clues to the listener about the nature of the speech that is important for recognition, such that listeners may use a closed response set to direct the focus of their attention in a manner similar to the results of training. For example, if the possible responses in a particular trial or a closed response set test are all identical except for their initial consonant, then listeners may choose to direct their attention primarily toward information at the beginning of the spoken word, ignoring the rest of the word in a manner quite unlike the distribution of attention when listening to speech normally. Similarly, if all of the possible responses in a particular trial differ only in terms of the manner of articulation of particular consonants (e.g., *fence* vs. *pence*), then listeners may choose to attend only to those aspects of the acoustic signal that distinguish fricatives from stops (such as rate of initial amplitude increase). Support for this hypothesis is provided by a study by Logan, Greene and Pisoni (1989). Even when scores are corrected for guessing, listeners tested using the standard closed-response version of the Modified Rhyme Test (see below) performed better than listeners taking the same test in an open-response format, suggesting that the set of available responses may help subjects distribute their attention more appropriately for recognition.

4.   COMMON TESTS OF SPEECH QUALITY

Using different tests to evaluate specific aspects of a TTS can provide a more thorough and detailed picture of what can be changed in order to improve it. There are a number of different reasons why one might be interested in testing the quality of synthetic speech, and, depending on these reasons, different tests or classes of tests may be preferred. For example, developers of TTS systems interested in identifying possible shortcomings in their systems may find intelligibility testing most useful, especially early in development. Initially, it may be most important to produce a system that can produce isolated words intelligibly, and therefore tests of segmental intelligibility might be most significant. As the system under development improves, special emphasis might be directed toward identifying and correcting problems in the calculation and realization of prosody, in order to improve the production of connected speech. However, during this phase, it might still be worthwhile to monitor segmental intelligibility, so that changes made to improve sentential intelligibility do not detract from segmental performance. Similarly, in developing a prosthetic voice for a particular user, special consideration might be given to identifying those ways in which it is or is not particularly natural, or does not sound like that talker would expect (or be expected) to sound (e.g., selecting appropriate pitch and formant frequencies for children's voices).

In contrast to these development applications, some tests are more useful for comparing existing products. For example, in setting up a computer for a blind user, it may be necessary to purchase one of many existing TTS systems. In this case, information about the rank ordering of these applications on one or more standard tests of acceptability may be quite useful. While the intelligibility and naturalness of a TTS system are obviously important for this application, the fact that both contribute strongly to overall measures of acceptability makes acceptability testing a more efficient method for assessing quality in this case.

Ideally, we would now list the names of existing standardized tests that investigate each of the factors discussed thus far. However, although the significance of all of those factors is relatively uncontroversial, existing tests take few of them into account. Although we will discuss tests for acceptability, intelligibility and naturalness in this section, almost all of the existing standardized tests are designed to evaluate segmental intelligibility, and therefore these dominate our discussion. In contrast, acceptability tests are more specific to each particular application being tested and often are highly constrained by the nature of the system being tested. In contrast, tests for naturalness or the realization of prosody are less common because these factors have only recently been identified as important to the synthesis of high quality speech. Finally, we will provide some general guidelines for the development and implementation of such tests.

4.1   Acceptability Tests

Although measures of acceptability are generally useful for ranking or otherwise comparing the quality of synthetic speech, there are not very many standard tests of acceptability available. In part, this is due to the strong degree of correlation between acceptability and intelligibility -- why measure acceptability if, during development, intelligibility must be measured anyway? Also, because acceptability may be strongly affected by factors related more to the application context than to the quality of the speech used in that context (as we have discussed in the sections on listening conditions and the text), it is possible that most tests of acceptability are designed for particular applications on an as-needed basis.

Acceptability tests do provide a fast overall measure of quality, which may be sufficient for many purposes. Since acceptability ratings reflect listeners' average opinions rather than detailed performance analyses, it is possible to rank systems on overall "goodness" of performance without specifically measuring intelligibility or naturalness. However, because they are so general, acceptability tests cannot diagnose specific problems with the TTS system, and they cannot help guide improvements during development. They are most usefully employed to rank synthesizers in terms of suitability, as judged by possible users, for use in a particular application. However, an examination of two standard tests of acceptability that have been described in print is informative for identifying important factors in the development of new tests of acceptability.

Mean Opinion Score (MOS). This is actually more of a general procedure than a specific test (Schmidt-Nielsen, 1995), and it serves to introduce the basic aspects of speech quality tests. In this test, sentences are presented auditorily to a large number of untrained listeners who rate the quality of the speech on a simple scale. Each scale consists of five discrete levels of quality, ranging from *excellent*, through *good*, *fair*, and *poor*, to *bad*. Scores are typically averaged across listeners and across tokens (and across talkers, if multiple "voices" from the same TTS system are used). Because the scale is so simple and typically listeners are not trained in making judgments about speech, a very large number of listeners is required to insure that the resulting score is not biased. Because there is no way of assuring that different groups of listeners necessarily use this scale in exactly the same way, there is the chance of serious variability in ratings of the same system by different groups of listeners. Therefore, in order to compare reliably multiple different TTS systems using the MOS, it is advisable to test all systems with the same group of listeners during the same testing session and use the rank-ordered scores, rather than the absolute scores, as the final assessment of (relative) quality.

Diagnostic Acceptability Measure (DAM). In contrast to MOS tests, the DAM (Schmidt-Nielsen, 1995) is a standardized test. Listeners are carefully screened for normal hearing, ability to discriminate speech sounds, and consistency in using rating scales. During testing, listeners' responses are recalibrated against known reference signals, and anchor and probe tokens are included in the set of stimuli in order to identify test-session-specific response biases on the part of individual listeners. Unfortunately, in its official form the DAM is not even applicable to the testing of TTS systems. The DAM is used primarily for the testing of speech coding and transmission systems, and therefore it uses a pre-recorded set of sentences spoken by six specific, "calibrated" human talkers. However, many of the concepts incorporated in its design are applicable to the design of tests of TTS systems as well. For example, the DAM uses 21 different rating scales, including ratings of overall speech quality, background (channel) noise characteristics, and general signal quality. Obviously, in most circumstances of testing a TTS system only those scales measuring speech quality are really of interest. However, the concept of breaking down the measurement of acceptability into sub-measurements of, for example, acceptability of pitch range, is very useful. In a sense, that is what is done by testing intelligibility and naturalness -- two factors which correlate strongly with overall ratings of acceptability.

4.2   Intelligibility Tests

Until relatively recently the biggest problem with TTS synthesis was intelligibility. Obviously, if a system cannot be understood, there's very little sense in trying to make it more natural sounding, or more socially appropriate for a given application. In addition, however, until recently most theories of speech perception, and therefore all of the applications that developed out of them, have been primarily "bottom-up." That is, they have been based on the assumption that, in understanding spoken language, human listeners first recognize phonemes, then combine those into words, and words into sentences, etc. Therefore, it was thought that it was most important to provide clearly intelligible phoneme synthesis, and understanding would follow from that. Because of this view, most existing tests are designed for evaluating segmental intelligibility in various contexts. In this section we first discuss some general issues in intelligibility testing, and then describe some of the most commonly used tests of intelligibility.

General Testing Issues. Most intelligibility testing tasks involve asking listeners to identify sounds they hear, either by transcribing what they hear into text, or by choosing labels (typically words or letters) from a list. Tests are usually scored according to the number of tokens correctly identified by listeners. Within this general framework, however, intelligibility tests differ in a number of ways. The set of possible responses can be open, meaning that listeners are free to identify tokens by choosing from the set of all possible words or phonemes that they know, or it can be closed, meaning that listeners must choose a response from among a set of possible responses provided as part of the test.

When using a closed response set, it is important to adjust the calculation of percent of tokens correctly recognized according to the probability that a token might be guessed correctly. This can be calculated using the following formula (ANSI, 1989):

$$R_A = R - W (n - 1)$$

In which R_A represents the adjusted percent correct score (the measure of intelligibility), R represents the number of items recognized correctly, W is the number of items identified incorrectly, and n is the total number of possible answers in each response set.

When time constraints or the proprietary status of a TTS system make it impossible to further improve intelligibility, it may be useful to consider the results of diagnostic testing in light of the intended use of the TTS system. For example, some TTS systems are intended to provide feedback or auxiliary information about written text or visual displays to users who can see the text or display it, if necessary (but where the use of exclusively visual displays is undesirable, such as in systems designed to be used by pilots of moving vehicles). Other systems, however, are designed to read text to the blind, or to provide speech for people who cannot otherwise communicate -- domains in which the speech cannot easily be supplemented or backed up with visual information. Clearly, the minimum acceptable level of speech quality might be much lower for systems of the first sort (where misunderstandings can, if necessary, be resolved by looking at a display) than for the second type of application where the speech must stand alone.

Phonetically Balanced (PB) Words. The Harvard phonetically (or, more accurately, phonemically) balanced (PB) word test was first proposed by Egan (1948) as part of a general hearing test. The PB word test consists of 1000 monosyllabic English words, in 20 lists of 50 words each. All of the words have a consonant-vowel-consonant (CVC) pattern, and the list is "balanced" in that, over the entire 1000 words, each phoneme appears with approximately the same frequency as it is expected to appear in spoken English. In testing, each word is generally presented in a neutral carrier sentence such as "Would you write _____ now" (ANSI, 1989). The test was designed as a pseudo-open-response test, meaning that listeners are first familiarized with all of the 1000 words in the list (usually by hearing them each several times prior to testing). However, in our lab we have had considerable success in using untrained college undergraduates as test listeners and presenting words in isolation. Using untrained listeners generally results in a higher degree of variability, both within and between tests, but that can be at least partially compensated for by increasing the number of listeners tested.

This test is the oldest test in the ANSI standard battery of intelligibility tests (ANSI, 1989), and according to Schmidt-Nielsen (1995) it has to some extent been superseded by closed-response tests. As we will discuss at the end of this section, there are advantages and disadvantages to both closed and open response tests. However, the set of words collected by Egan (1948) remains useful for both.

Modified Rhyme Test (MRT). The MRT was developed initially by House and his colleagues (House, Williams, Hecker and Kryter, 1965) in an effort to improve the efficiency of intelligibility testing. It is based on Fairbanks (1956) Rhyme Test, but is designed to be easier to administer and score than the original Rhyme Test, which does not appear to be used anymore. The MRT is made up of 50 lists containing six words each (300 words total). Almost all words in all lists are monosyllabic, with a CVC structure, though there are some lists in which the words are CVCC, such as *must, vest,* and *hark.* Within each list words differ only in terms of their initial consonant or final consonant. Each trial consists of the visual presentation of all six words from a single list, in conjunction with the auditory presentation of one of the six words from the list. Words may be presented either in isolation or in a neutral carrier sentence (as for the PB word lists). The order of presentation of the 50 lists can be randomized, as should be the choice of which of the six words in each list is spoken and the physical location of the target word among the alternative words displayed (in order to avoid position bias in response choices). Because this is a closed response set test, listeners have a measurable chance of getting the correct answer simply by guessing, and any scores derived by this test must be adjusted for guessing using the formula discussed above in the section on message set.

Diagnostic Rhyme Test (DRT). The DRT (Voiers, 1977; Voiers, 1983) is another variant of the Rhyme Test (Fairbanks, 1956). It is designed to provide specific information about the degree to which a speech system succeeds at producing or preserving a distinction between phonemes that differ according to particular distinctive features (Voiers, 1977). Like the MRT, the DRT is a closed response set test using monosyllabic, (mostly) CVC English words. The 192 words of the DRT are organized in 6 lists of 16 pairs each. Within each pair the words differ only in their initial consonant. The differences between the consonants in a given pair can be characterized as differences in the value of a particular feature, using a set of six distinctive features developed by Voiers (1977), based on work by Jakobsen, Fant and Halle (1955). These features are *voicing, nasality, sustention, sibilation, graveness,* and *compactness.* For example, in the words *veal* and *feel* the consonants [v] and [f] differ only according to their *voicing.* Similarly, in the pair *fence* and *pence* the initial consonants differ only in terms of their *sustention.* For a more detailed discussion of the nature and use of these features, please see Voiers (1977).

In performing the DRT, words are presented visually in pairs prior to the auditory presentation of one of the two words (presented in isolation). Listeners must choose between the two words of the pair to make their identification. Thus, the probability of making a correct guess is 50%. For presentation, pairs should be randomly chosen from the set of all pairs, rather than being blocked by feature, otherwise listeners may easily deduce which feature is significant for each list, and direct their attention primarily to the appropriate distinctions (see the section on attention and learning).

Consonant Identification (CI). In the course of explicitly evaluating the role of response-set size on tests of intelligibility, Greenspan, Syrdal, and Bennett (in press) describe a test of consonant intelligibility that, in contrast to the closed-response set DRT, is specifically open-response. The test of Consonant Intelligibility (CI) that they describe is not (yet) a standard test, but the techniques it employs have been used in one form or another in intelligibility testing since the beginning of this century (Greenspan et al., in press).

The CI uses syllables, not words, as stimuli. Each syllable consists of a single consonant followed by a single vowel (so-called CV syllables). There are a total of 21 consonants (all of the consonants which can appear utterance-initially in English except for the voiced /th/ as in *the*, *this*, or *that*). Each of these consonants is produced in conjunction with each of the three "point" vowels, /i/, /a/, and /u/ (as in *heat*, *hot*, and *hoot*). These CV syllables are presented to listeners blocked by vowel, so that the only difference between syllables within a block is their initial consonant. Syllables are presented in pseudo-random order within a block (constrained so that two syllables starting with the same consonant cannot appear consecutively), and blocks may be presented in random order as well. Listeners are told that each sound they hear will be a CV syllable, and they are asked to write down the initial consonant of each syllable they hear. Throughout the experiment a list of all the possible consonants is visible, and the listeners are familiarized with the orthographic code for representing each sound before the experiment begins.

One principle disadvantage to this testing method is that it is very limited in terms of the linguistic-phonetic context it tests. It only tests intelligibility of consonants in utterance-initial, pre-vocalic position in stressed monosyllables. However, it is clearly possible to design test stimuli which test consonants (and even vowels) in a similar manner within a more varied set of contexts. For example, Spiegel, Altom, Macchi, and Wallace (1990) describe a corpus of monosyllabic words and pseudo-words for testing the intelligibility of consonants in a wider range of contexts than is provided by either the Rhyme tests or CI tests (including pre- and post-vocallically and within clusters). This corpus also has more diagnostic application than the PB word test.

Polysyllabic And Polymorphemic Word Tests. For the most part, intelligibility testing has been carried out using almost exclusively monomorphemic, monosyllabic test corpora. Such words are shorter and less complex than many words that might be expected to be produced by a TTS system. Because they are short and simple, monomorphemic words offer little opportunity for listeners to use lexical knowledge to aid in recognition. It is well known that meaningful contexts provide a great deal of aid in recognizing spoken words, and longer words with complex morphology may provide listeners with more contextual cues for phoneme identification than do shorter monomorphemic words. If this is the case, estimates of intelligibility based on monosyllabic materials might underestimate the intelligibility of multisyllabic, multimorphemic words produced by particular speech synthesizers. Alternatively, the rules which govern the correct production of longer, more morphologically complex words may not be as well understood, or as easily incorporated into a TTS system, and therefore intelligibility ratings based on monosyllabic, monomorphemic corpora may *overestimate* the intelligibility of the system. In short, although a TTS system will likely be expected to produce long and morphologically complex words intelligibly, monomorphemic, monosyllabic test corpora simply cannot reveal any information relevant to assessing the intelligibility of morphologically complex speech.

However, by following the guidelines established by existing tests, it is possible to develop test corpora consisting of morphologically complex multisyllabic words, and to use these corpora to compare the intelligibility of TTS systems (Francis and Nusbaum, in press). The major principles involved are to minimize the role that other cues to word identity play. Such tests can be constructed using sets of words which are roughly matched for segmental characteristics, and differ primarily in terms of morphological complexity. For example, the words *vest, vestment, vestmented, versus,* and *vesicle* are all roughly matched in terms of stress pattern (each has primary stress on the initial syllable) and phonological characteristics (all start with /v/, all but one share the same stressed vowel, all have an /s/ in the cluster ending the stressed syllable, etc.). In this case the use of real words, made necessary by the desire to test morphological aspects of TTS synthesis, precludes the incorporation of pseudo-words to create a truly balanced set. However, for testing other lexical phenomena such as word-level stress for which word meaning is less significant, more segmentally matched words and pseudo-words could also be developed. Alternatively, if trained listeners are to be used, it is conceivable to use perfectly matched sets of words and pseudo-words, and simply train listeners to recognize these words until they reach a predetermined response-time criterion on all test items.

Sentence Tests. Just as word-level linguistic characteristics like morphology and prosody contribute to intelligibility and should be tested, the sentence in which a word or segment appears can also provide significant cues to its identity. It is well known that the acoustic patterns of words extracted from fluent speech are less intelligible than words produced in isolation. However, words heard in sentence context are generally more intelligible than words produced in isolation because sentences provide prosodic, syntactic, and semantic cues to word identity (Miller and Isard, 1963). If a TTS system is to be expected to produce complete sentences, then its ability to produce intelligible sentences, and intelligible words and segments embedded in real sentences, should also be tested. Although there are again no truly standard tests for sentential intelligibility, there do exist a number of useful sentence lists that can be incorporated into different kinds of intelligibility testing. Note that we are not talking here about frame sentences such as are used in the PB word test of word intelligibility. Although these provide prosodic, semantic and syntactic cues appropriate to identifying that a particular token is a word and should be treated as such, they are specifically designed *not* to provide any such cues with respect to the *identity* of the target word.

The two most well-known sets of sentences are the Harvard sentences, developed by Egan (1948) as part of the same hearing test as the PB word list, and the Haskins sentences (Nye and Gaitenby, 1973). The Harvard sentences were designed to be meaningful sentences that are not particularly predictable. That is, although upon hearing any particular sentence it is easy to understand what the sentence means, hearing only a part of the sentence will not give much clue as to the content of the rest of the sentence. For example, the sentence *These days a chicken leg is a rare dish* is clearly meaningful. However, given any fraction of the sentence such as *These days* or *a chicken leg is* or *is a rare dish*, the rest of the sentence is not predictable (in contrast to the case of a highly predictable sentence such as *He likes his coffee with cream and sugar*, where one could easily predict *with cream and sugar* following *He likes his coffee*). The intelligibility of Harvard sentences is usually calculated in terms of the percentage of key content words (usually out of five) correctly recognized or transcribed. Thus, the Harvard sentences are useful for evaluating the ability of a TTS system to produce segment, word and sentence level information in combination, while controlling for the contributions of real-world and semantic knowledge.

The Haskins sentences, in contrast, were designed to eliminate any role of semantic or real-world knowledge in word recognition. Like the Harvard sentences, the Haskins sentences are scored according to the number of content words recognized correctly. However, unlike the Harvard sentences, the Haskins sentences are semantically more or less uninterpretable. As with all such sentences, it is certainly possible to imagine some possible world in which this sentence makes sense. However, it seems reasonable to assume that such conceptual gymnastics are performed only *after* the sentence has been recognized, and that knowledge of such ad hoc hypothetical worlds does not usually contribute significantly to the recognition of spoken language. The important point is that these sentences are, grammatically, completely acceptable. Thus, if a listener is able to correctly recognize the content words in the sentence *The old corn cost the blood*, they must have done so purely on the basis of the acoustic characteristics of the signal and their knowledge of speech acoustics, morphology and syntax. The listener's knowledge of word co-occurence, word meaning, and the real world cannot provide information sufficient to transcribe all of the content words in this sentence if only some of them are recognized.

A third set of sentences for evaluating speech quality of coded speech has been developed by Huggins and Nickerson (1985). Although these sentences, called phoneme-specific sentences, have not yet been as carefully tested and normed as the words used in the DRT, they provide a similar kind of information about the ability of a synthesizer or coding scheme to produce particular segments intelligibly. However, unlike the DRT, the phoneme-specific sentence test evaluates segmental intelligibility in the context of a task (transcribing sentences) that is somewhat closer in nature to the usual task involved in using a TTS synthesizer. Phoneme-specific sentences are sentences that contain only particular classes of phonemes, or emphasize particular classes. For example, the sentence *Where were you a year ago* is constructed out of only vowels and vowel-like segments.[1] Similarly, all of the consonants in the sentence *Which tea party did judge Baker go to* are stops or affricates. The first sentence, and ones similar to it, would be useful for testing the quality of the contributions of formant and formant-related aspects of synthesis to intelligibility. The second sentence, and ones like it, would be well suited to evaluating the quality of the synthesis of noisy consonants and abrupt spectral transitions.

The test described by Huggins and Nickerson (1985) was intended as an acceptability-rating task, rather than as a test of intelligibility. In that experiment listeners were asked to perform two tasks. The first involved rank-ordering identical sentences produced with different coding schemes. The second involved explicit rating of the quality of each of the sentences used in the ranking task. These data are then amenable to many different kinds of subsequent analyses, including but not limited to the multi-dimensional scaling techniques used in the original paper. If the performance task involved recognition or transcription of these sentences, rather than quality evaluation, the results of such a test would be useful as a rough estimate of segmental intelligibility.

Comprehension Tests. There are many other tests and classes of tests that have been developed for the evaluation of the comprehensability of speech (as opposed to merely the intelligibility), but most of them are not well suited for the evaluation of TTS synthesis (see Schmidt-Nielsen (1995) for a more thorough discussion). Many of the tests she discusses are designed to test the quality of communication systems such as telephones, and therefore require two participants to interact over the same channel. Thus, they are not well suited to evaluating speech quality when one of the interlocutors is a machine. However, the kinds of tasks involved (problem solving, game playing) might be adaptable with some imagination to testing TTS systems.

The most important factor in a test of comprehension is that listeners must have to use the meaning of the message in some objective, quantifiable way. For example, many comprehension tests involve asking users to answer questions or follow directions, such that the subjects' performance in terms of number of questions answered correctly, or number of tasks correctly carried out can be easily calculated. In such tests it is important that the questions to be answered or the tasks to be carried out be simple enough that subjects' inability to perform the task or answer the question does not affect the overall score. On the other hand, the tasks and questions should not be so predictable that subjects can achieve a score above chance without really understanding the speech. Schmidt-Nielsen (1985) describes an excellent test, called the Naval Research Laboratory (NRL) test that involves playing a game based on the traditional graph-paper and pencil game. The test, as currently implemented, is ideal for testing a two-way communication system such as a telephone or radio, but could be adapted for use with synthetic speech by replacing one of the players with a TTS synthesizer. Either one could be controlled in real time by a human experimenter or, possibly, even a completely artificial system including a speech recognition engine and a PC running an actual Battleship game program. In this game, the object is to sink the opponent's ships by correctly guessing which squares the ships lie in on a five-by-five grid. Squares are guessed by calling them out in terms of a row-and-column grid system. Traditionally, rows are labeled alphabetically (A, B, C, D, E. or *alpha, bravo, charlie, delta, echo*) and columns are labeled with numbers (1, 2, 3, 4, 5). In this format, the task is ideal for testing the ability of a TTS system to generate a small set of words (the digits from 1 to 5, and the letters from A to E, or from *alpha* to *echo*). However, it is conceivable that the test could be modified to include any small set of words, including, for example, those used in the MRT. A significant advantage of this task over other kinds of tasks is that it provides a natural and familiar context for spoken communication, rather than the highly formal testing paradigms used in many of the other tests discussed here.

4.3   Naturalness Tests

Currently there are no standard tests of naturalness. This lack may partly be due to the problem that naturalness has not yet been defined objectively or even operationalized in terms of a characteristic of human performance. In contrast, intelligibility can be defined as "percent correct words/segments/sentences identified or transcribed," and acceptability can be operationalized as "good, bad, or ugly" to listeners asked to rate quality. The solution to this lack of an operational definition of naturalness has two parts. First, we must identify features that are likely candidates for cueing the relative humanness of speech. Secondly, we must decide how to ask listeners which speech they consider natural. Nusbaum, Francis, and Henly (1995) describe two experiments that serve to illustrate the design and implementation of naturalness tests.

Identifying Features. First, we do know something about what aspects of speech contribute to the perception of naturalness, and we can develop tests of the quality of these features. As discussed earlier in this chapter, a great deal of speech research had been devoted to discovering how information is encoded in human speech. Although much of this work has concentrated on linguistic factors, there is also a sizable body of literature on other factors (e.g. Klatt and Klatt, 1990; Laver, 1980). Since TTS synthesis necessarily involves choosing which of these features to include in a synthetic voice, it is a small matter to select these same features for testing.

This task is made slightly more complicated by the observation that, because many of the characteristics of speech that contribute to naturalness also contribute to intelligibility, it is important to eliminate the role of intelligibility in any task used for assessing naturalness. The differences in intelligibility between natural and synthetic speech are still sufficiently great that they affect judgments of naturalness and acceptability. Therefore, in order to test the role of non-linguistic aspects of speech quality independently from questions of (linguistic) intelligibility, it is necessary to eliminate all linguistic components of the task used in evaluation. Even if the test instructions simply ask subjects to rate the quality or naturalness of an utterance, listeners' responses will reflect not only the affective quality of the speech, but also its intelligibility .

One way of eliminating the role of intelligibility is to make all tokens equally intelligible. In practice, the simplest way of doing this involves making all tokens (close to) 100% intelligible, or making them all completely unintelligible.[2] In a recent series of experiments in our lab, both methods were employed with success (Nusbaum, et al., 1995). The first method we used involved telling listeners the linguistic identity of the tokens they heard, thereby insuring that all tokens were intelligible. Since the naturalness judgment task we used did not involve evaluating listeners' ability to *identify* the stimuli, the ability or inability to extract this information from the signal should not have influenced listeners' judgments of naturalness. It is, however, possible that listeners' opinions about the degree to which each synthesizer *achieved* the identified target phoneme still could have affected naturalness judgments. In a second test we aimed for 0% intelligibility, and attempted to remove all cues to segmental phonetic information. By low-pass filtering the speech signals at 200 Hz, we insured that listeners would have to make their naturalness judgments without regard for whether they could understand what was being said.

Defining Naturalness. Following the model of acceptability testing, we can define naturalness *a priori* as "human-like" and structure our tests to ask listeners to specifically rate how human a particular signal is, or to rank utterances according to their judgments of how human the talker sounds. If these tests also include natural human speech as a reference against which to compare synthetic speech we will then also be able to assess the degree to which the particular test is capable of providing accurate information about the naturalness of synthetic speech. Rather than asking for overt judgments of naturalness, we could also measure naturalness in terms of the probability that listeners are likely to consider a talker to be a natural talker. The more natural sounding a voice is, the more likely listeners should be to classify it as a human voice. In our lab we use a two-alternative forced-choice task in which subjects classify stimuli as either HUMAN or COMPUTER. From these results, we can compute the probability that any particular stimulus would be rated as human or computer, and compare talkers on the basis of these scores.

Since natural human speech should, by definition, be evaluated as completely, or almost completely natural,[3] any test of naturalness on which human speech fails to score sufficiently high should be suspect. For example, in the first experiment described in Nusbaum et al. (1995) we discovered that in the context of the vowel /u/ listeners rated all talkers as almost uniformly unnatural. Listeners were able to reliably distinguish between human and synthetic talkers overall, and in other vowel contexts. This suggests that the method used to create our stimuli resulted in an /u/ that sounded sufficiently unnatural (perhaps due to the lack of any diphthongal quality) that voice quality differences were obscured.

5. THE ROLE OF BASIC SPEECH RESEARCH

For the most part, TTS systems have been developed by attempting to replicate, as closely as possible, the production capabilities of human talkers. Especially with respect to issues of intelligibility and naturalness, the primary method of development has been to identify acoustic patterns corresponding to linguistic features through the careful analysis of samples of natural human speech. Then, these characteristics are replicated as accurately as possible in the synthetic speech (cf. Klatt and Klatt, (1990) for an excellent example of this method). Evaluation techniques have, thus far, been used primarily in a confirmatory manner, to insure that the acoustic patterns modeled on spectrograms and waveform analyses of human speech are sufficiently intelligible, acceptable or natural for particular applications.

We argue that, besides understanding how talkers produce speech, the development of high quality synthetic speech also requires an understanding of what listeners need and expect from speech in the context of interacting with a TTS synthesizer. Listeners are not merely passive signal acquisition devices, but rather active participants in the process of understanding speech (Nusbaum and Schwab, 1986; Nusbaum and Magnuson, 1996). Thus, the abilities of comprehending spoken language and making judgments of the acceptability and naturalness of spoken language are simultaneously constrained and facilitated by the interaction of the structure of the text and the demands of the task. While knowledge of language and task domain can facilitate comprehension by an active listener despite an impoverished speech signal, the attention and memory demands of complex listening tasks and the insufficient or misleading cue structure of synthetic speech can also prove more detrimental to comprehension by an active listener. Ultimately, the development and application of tests for evaluating the quality of speech produced by a TTS system must reflect not only an understanding of speech production, but also an understanding of human speech perception.

ACKNOWLEDGMENTS

We would like to thank Dr. Astrid Schmidt-Nielsen for helpful suggestions on locating technical documents.

REFERENCES

Allen, J., Hunnicutt, S., Carlson, R., and Granstrom, R. (1979). MITalk-79: The MIT text-to-speech system. *Journal of the Acoustical Society of America*, 65 (suppl. 1), S130.

American National Standards Institute. (1989). *Method for Measuring the Intelligibility of speech Over Communication Systems* (ANSI S3.2-1989 R1995 / Acoustical Society of America Catalog No. 85-1989). New York: Acoustical Society of America.

Baddeley, A. D., and Hitch, G. J. (1974). Working memory. In G. Bower (Ed.), *Recent advances in learning and motivation* (Vol. VIII). New York: Academic Press.

Baldwin, K., Francis, A. L., and Nusbaum, H. C. (submitted). Learning to listen: The effects of training on attention to acoustic cues.

Blumstein, S. E., and Stevens, K. N. (1981). Phonetic features and acoustic invariance in speech. *Cognition, 10,* 25-32.

Cooper, F. S., Delattre, P. C., Liberman, A. M., Borst, J. M., and Gerstman, L. J. (1952). Some experiments on the perception of synthetic speech sounds. *Journal of the Acoustical Society of America, 24,* 597-606.

Delattre, P. C., Liberman, A. M., and Cooper, F. S. (1955). Acoustic loci and transitional cues for consonants. *Journal of the Acoustical Society of America, 27,* 769-773.

Denes, P. (1955). Effect of duration on the perception of voicing. *Journal of the Acoustical Society of America, 27,* 761-764.

Dorman, M. F., Studdert-Kennedy, M., and Raphael, L. J. (1977). Stop-consonant recognition: Release bursts and formant transitions as functionally equivalent, context-dependent cues. *Perception and Psychophysics*, *22*, 109-122.

Egan, J. P. (1948). Articulation testing. *Laryngoscope*, *58*, 955-991.

Fairbanks, G. (1956). Test of phonemic differentiation: The rhyme test. *Journal of the Acoustical Society of America*, *30*, 596-600.

Fant, G. (1991). What can basic research contribute to speech synthesis? *Journal of Phonetics*, *19*, 75-90.

Fitch, H. L., Halwes, T., Erickson, D. M., and Liberman, A. M. (1980). Perceptual equivalence of two acoustic cues for stop-consonant manner. *Perception and Psychophysics*, *27*, 343-350.

Forster, K. I. (1976). Accessing the mental lexicon. In R. J. Wales and E. Walker (Eds.), *New approaches to language mechanisms*. (pp. 257-287). Amsterdam: North Holland.

Francis, A. L., and Nusbaum, H. C. (in press). The effect of lexical complexity on segmental intelligibility. To appear in *International Journal of Speech Technology*.

Ganong, W. (1980). Phonetic categorization in auditory word perception. *Journal of Experimental Psychology: Human Perception and Performance*, *6*, 110-125.

Gardner-Bonneau, D. (1998). Personal communication.

Greene, B. G., and Pisoni, D. B. (1988). Perception of synthetic speech by adults and children: Research on processing voice output from text-to-speech systems. In L. E. Bernstein (Ed.), *The vocally impaired: clinical practice and research*. Philadelphia: Grune & Stratton.

Greenspan, S. L., Syrdal, A. K., and Bennett, R. W. (in press). An evaluation of the diagnostic rhyme test. To appear in *International Journal of Speech Technology*.

Greenspan, S. L., Nusbaum, H. C., and Pisoni, D. B. (1988). Perception of synthetic speech produced by rule: Intelligibility of eight text-to-speech systems. *Behavioral Research Methods, Instruments, and Computers*, *18*, 100-107.

Hecker, M. H., and Williams, C. E. (1966). Choice of reference conditions for speech preference tests. *Journal of the Acoustical Society of America*, *39*, 946-952.

House, A. S., Williams, C. E., Hecker, M. H. L., and Kryter, K D. (1965). Articulation testing methods: Consonantal differences with a closed response set. *Journal of the Acoustical Society of America*, *37*, 158-166.

Huggins, A. W. F., and Nickerson, R. S. (1985). Speech quality evaluation using "phonemic-specific" sentences. *Journal of the Acoustical Society of America*, *77*, 1896-1906.

Hunnicutt, S. (1995). The development of text-to-speech technology for use in communication aids. In A. Syrdal, R. Bennett, and S. Greenspan (Eds.), *Applied Speech Technology*. (pp. 547-563). Boca Raton: CRC Press.

Jakobsen, R., Fant, C. G. M., and Halle. M. (1955). *Preliminaries to speech analysis: The distinctive features and their correlates* (MIT Acoustics Laboratory Technical Report No. 13). Cambridge, MA: Acoustics Laboratory.

Klatt, D. H. (1976). Linguistic uses of segmental duration in English: Acoustic and perceptual evidence. *Journal of the Acoustical Society of America*, *59*, 1208-1221.

Klatt, D. H. (1980). Software for a cascade/parallel formant synthesizer. *Journal of the Acoustical Society of America, 67*, 971-995.

Klatt, D. H., and Klatt, L. C. (1990). Analysis, synthesis and perception of voice quality variations among female and male talkers. *Journal of the Acoustical Society of America, 87*, 820-857.

Laver, J. (1980). *The phonetic description of voice quality.* Cambridge: Cambridge University Press.

Lee, L., and Nusbaum, H. C. (1989, May). *The Effects of Perceptual Learning on Capacity Demands for Recognizing Synthetic Speech.* Paper presented at the Acoustical Society of America, Syracuse.

Liberman, A. M., Cooper, F. S., Shankweiler, D. P., and Studdert-Kennedy, M. (1967). Perception of the speech code. *Psychological Review, 74*, 431-461.

Lisker, L. (1978). Rabid vs. rabid: A catalogue of acoustic features that may cue the distinction. *Haskins Laboratories Status Report on Speech Research, SR-54*, 127-132. New Haven: Haskins Laboratories.

Lisker, L. (1975). Is it VOT or a first-formant transition detector? *Journal of the Acoustical Society of America, 57*, 1547-1551.

Logan, G. D. (1979). On the use of a concurrent memory load to measure attention and automaticity. *Journal of Experimental Psychology: Human Perception and Performance, 5*, 189-207.

Logan, J. S., Greene, B. G., and Pisoni, D. B. (1989). Segmental intelligibility of synthetic speech produced by rule. *Journal of the Acoustical Society of America, 86*, 566-581.

Luce, P. A., Feustel, T. C., and Pisoni, D. B. (1983). Capacity demands in short-term memory for synthetic and natural words lists. *Human Factors, 25*, 17-32.

Miller, G. A., and Isard, S. (1963). Some perceptual consequences of linguistic rules. *Journal of Verbal Learning and Verbal Behavior, 2*, 217-228.

Morton, J. (1969). Word recognition. In J. Morton and J. D. Marshall (Eds.), *Psycholinguistics 2: Structures and processes* (pp. 107-156). Cambridge, MA: The MIT Press.

Nusbaum, H. C., Francis, A. L., and Henly, A. S. (1995). Measuring the naturalness of synthetic speech. *International Journal of Speech Technology*, 1, 7-19.

Nusbaum, H. C., and Magnuson, J. S. (1996). Talker normalization: Phonetic constancy as a cognitive process. In K. A. Johnson and J. W. Mullennix (Eds.), *Talker variability and speech rocessing* (pp. 109-132). New York: Academic Press.

Nusbaum, H. C., and Pisoni, D. B. (1985). Constraints on the perception of synthetic speech generated by rule. *Behavior Research Methods, Instruments, & Computers, 17*, 235-242.

Nusbaum, H. C., and Schwab, E. C. (1986). The role of attention and active processing in speech perception. In E. C. Schwab and H. L. Nusbaum (Eds.), *Pattern recognition by humans and machines: Speech perception, Vol. 1* (pp. 113-157). New York: Academic Press.

Nye, P. W., and Gaitenby, J. (1973). Consonant intelligibility in synthetic speech and in a natural control (modified rhyme test results). *Haskins Laboratories Status Report on Speech Research, SR-33*, 77-91. New Haven: Haskins Laboratories.

Olive, J. P., Greenwood, A., and Coleman, J. (1993). *Acoustics of American English speech.* New York: Springer Verlag.

Pickett, J. M. (1980). *The sounds of speech communication.* Boston: Allyn and Bacon.

Pisoni, D. B., Nusbaum, H. C., Luce, P. A., and Slowiaczek, L. M. (1984). Speech perception, word recognition, and the structure of the lexicon. *Speech Communication, 4,* 75-95.

Port, R. F., and Dalby, J. (1982). Consonant/vowel ratio as a cue for voicing in English. *Perception and Psychophysics, 32,* 141-152.

Ralston, J. V., Pisoni, D. B., and Mullennix, J. W. (1995). Perception and comprehension of speech. In A. K. Syrdal, R. W. Bennett, and S. L. Greenspan (Eds.), *Applied speech technology* (pp. 233-288). Boca Raton: CRC Press.

Repp, B. H. (1982). Phonetic trading relations and context effects: New experimental evidence for a speech mode of perception. *Psychological Review, 92,* 81-110.

Schmidt-Nielsen, A. (1985). Problems in evaluating the real-world usability of digital voice communication systems. *Behavior Research Methods, Instruments, and Computers, 17,* 226-234.

Schmidt-Nielsen, A. (1995). Intelligibility and acceptability testing for speech technology. In A. K. Syrdal, R. W. Bennett, and S. L. Greenspan (Eds.), *Applied speech technology* (pp. 195-232). Boca Raton: CRC Press.

Schwab, E. C., Nusbaum, H. C., and Pisoni, D. B. (1985). Effects of training on the perception of synthetic speech. *Human Factors, 27,* 395-408.

Spiegel, M. F., Altom, M. J., Macchi, M., and Wallace, K. L. (1990). Comprehensive assessment of the telephone intelligibility of synthesized and natural speech. *Speech Communication, 9,* 279-291.

Syrdal, A. K. (1989). Improved duration rules for text-to-speech synthesis. *Journal of the Acoustical Society of America, 85,* S43.

Voiers, W. D. (1983, Jan/Feb). Evaluating processed speech using the Diagnositc Rhyme Test. *Speech Technology,* 30-39.

Voiers, W. D. (1977). Diagnostic evaluation of speech intelligibility. In M. E. Hawley (Ed.), *Speech intelligibility and speaker recognition* (pp. 374-387). Stroudsburg, PA: Dowden, Hutchinson, and Ross.

Walley, A. C., and Carrell, T. D. (1983). Onset spectra and formant transitions in the adult's and child's perception of place of articulation in stop consonants. *Journal of the Acoustical Society of America,* 73, 1011-1022.

Williams, C. E., and Hecker, M. H. (1968). Relation between intelligibility score for four test methods and three types of speech distortion. *Journal of the Acoustical Society of America, 44,* 1002-1006.

[1] The phonemes /w/, /y/ and post-vocalic /r/ can be considered vowels for some purposes, because, like true vowels, their spectra can be generated without the use of zero poles, and do not change amplitude abruptly.

[2] Though any other arbitrary rating is equally useful, in practice this would likely be much harder to achieve.

[3] Of course, the absolute degree to which human speech is considered natural will in part depend on the specific processing performed on that speech for testing. For example, band-pass filtering and amplitude normalization, which are routinely done to make test stimuli more similar to one another, may cause speech to sound overall less natural. More extreme processing, such as severe low-pass filtering to remove segmental information, may have a more serious effect. However, as long as all speech samples, synthetic and natural, are processed in the same way, it may be assumed that, in the absence of counter-evidence, all processing will affect their naturalness ratings to the same degree. Note also that different unprocessed human talkers may vary in absolute naturalness ratings as well.

4 PHONOLOGICAL RULES FOR SPEECH SYNTHESIS

Michel Divay

Université de Rennes
Institut Universitaire de Technologie

Key words: Letter-To-Sound Rules, Grapheme-To-Phoneme Translation, Text Normalization, Rewriting Formalism

Abstract: *This chapter presents the grapheme-to-phoneme transcription, one of the first steps of speech synthesis from text. Algorithms have to process regular texts along with abbreviations, acronyms and digits. The text to synthesize is converted into phonemes that are then used to generate the speech signal. This grapheme-to-phoneme conversion is language dependent, and can be achieved using a rewriting rule formalism. Difficulties and solutions for both English and French are presented.*

1. INTRODUCTION

Reading a text aloud for a human requires the mental conversion of the text into a sequence of phonemes, and the pronunciation of those phonemes, adding, according to the meaning of the text, intonation for the entire sentence and pauses at the syntactic boundaries of groups. Each phoneme has a duration that depends on its position in the word, and on the position of that word in the sentence. The speaker is not aware of all the details of this process, which has been magically acquired in his or her infancy as far as the mother tongue is concerned.

An algorithm for speech synthesis must mimic explicitly all the details of each step (grapheme-to-phoneme conversion, phoneme duration, intonation of the sentence). These values are then sent to the signal synthesizer, which updates the different parameters of the synthesizer, and then produces the wave signal. This chapter focuses on grapheme-to-phoneme conversion only, explaining the different problems encountered during that step, and giving some examples of the way problems have been managed for both English and French.

Letter-to-sound programs are also used for other purposes: word or name lookups for database searches and phonetic lexicon creation for speech recognition.

Grapheme-to-phoneme conversion can be used to retrieve an item without knowing the right spelling for either words or proper names. First, a phonetic index is created with all the words or proper names to recognize. To retrieve a word or a proper name without knowing the exact spelling requires converting the grapheme string into phonemes, and searching for the phonetic string in the phonetic index. This is typically a database search based on a key that is a phonetic index. In the database system Taurus (Taurus, 1993), a phonetic key can be declared for any text field when describing the structure of the database. Another typical application would be to correct misspelled words using phonetics: when a word is not found in the grapheme index, the word is converted into phonetic symbols, and searched for in the phonetic string. One or several words can be suggested to replace the original spelling.

Grapheme-to-phoneme conversion is also used in speech recognition to generate the phonemes corresponding to the vocabulary of the application the system has to recognize. Several phoneme strings could be produced for a word to take into account the different pronunciations of that word.

Different methods have been implemented to achieve this conversion. The simplest one is to use a dictionary with the equivalent phoneme string for each entry. If the synthesizer is a general-purpose text-to-speech synthesizer, it has to convert any input string into speech, and this includes derived forms, new words, proper nouns, numbers, and acronyms. The dictionary would have to be very large and evolve constantly. But knowing the grapheme pronunciation of an isolated word is not always sufficient. The pronunciation sometimes depends on the word context (as *the apple* versus *the boy* in English). For French, a linking phoneme is often added between two words in some contexts.

To develop the most **general, independent and concise** set of rules, firstly, the rules have to be general to process the largest number of words, then, they have to be independent, and not interfere with one another, and finally, they have to be concisely written, and yet understandable. The development of a letter-to-sound set of rules amounts to teaching a computer how to read (pronounce) a language.

The conversion varies from one alphabetic language to the other, and depends on the fit between graphemes and corresponding phonemes. The conversion is very straightforward for a language like Spanish, but is complicated for English or French. Syllabary writings, as in Japanese, or logographic systems, as in Chinese, are not concerned with this grapheme-to-phoneme problem, but have other problems to cope with, for instance, Kanji reading to input the characters into the computer (Hakoda, Kitai, and Sagayama, 1997).

2.   THE HISTORIC EVOLUTION OF ENGLISH AND FRENCH

For both English and French, there is a large gap between the written and spoken components of the language, which have evolved at different rates. The pronunciation has constantly changed whereas the spelling has evolved more slowly, and has been stabilized by the publication of the first dictionaries. In some cases, letters have been added to words to reflect their Latin origins (*b* in doubt for English, *p* in *compter* for French), or to distinguish different homophones. In China, the same logographic characters can be pronounced differently in Beijing or Shangai, and are in that way similar to the difference of pronunciation between English and French for a word like *aggravation* with the same spelling and meaning but with a totally different pronunciation.

It is fairly difficult to construct letter-to-sound rules for the English language. The vocabulary is estimated to be about one million words including local jargon words, scientific and technical words. The vocabulary originates from both Germanic and Roman (Latin and French) and results from successive immigration waves. Many French words were introduced after the Norman invasion (1066 AD) resulting in the addition of phonemes (/f/-/v/ distinction). Between 1400-1550 AD, a shift in the pronunciation of vowels (Great Vowel Shift) occurred. It had no effect on spelling, but widened the gap between graphemes and phonemes: *kite* and *kit* used to be pronounced respectively with a long and short /i/; *name* and *face* were pronounced with phoneme /aː/ (Wells, 1982; Ben Crane, Yeager, and Whitman, 1981; CD-ROM Encarta, 1995).

French has evolved from Latin. During the Roman invasion, people began to speak Latin, and continued to do so after the Roman withdrawal in 476 AD. The language evolved in different ways as few people knew how to read. The first written text in the new language, called *Roman (*[romã]*),* only appeared in the 9th century. Then official papers usually written in Latin began to be written in *Roman*. Standardization in spelling had been vague until dictionaries, schools, and laws enforced a standard spelling in the 17th century (Catach, 1978). Nowadays, new words are still added to the language: foreign words (parking, weekend, pizza, spaghetti), technical words (ordinateur, cédérom), etc.

3.   THE COMPLEXITY OF THE CONVERSION FOR ENGLISH AND FRENCH

3.1   English

"Pronunciation of vowels and vowels digraphs, consonants and consonant clusters, prefixes and suffixes, is highly dependent upon context" (Hunnicut, 1976).

There are many examples showing the gap between written text and pronunciation. The phoneme [ʃ] for instance can be written with many spellings:

c	social, ocean	sch	schilling
ch	champagne	sh	shame, ship
chs	fuchsia	ss	fission, pressure
s	sugar	t	ambitious
sc	conscience	x	anxious

The grapheme *ch* can represent different phonemes:

[ʃ]	**ch**ampagne
[t][ʃ]	in**ch**
[k]	**ch**olera

The *ch* case is relatively simple. For vowels, the number of different pronunciations depending on the context is much larger. Vowel *a* between two consonants, for instance, can be represented by the following phonemes:

[ɔː]	bald, small	[ə]	saloon
[ɛə]	grammarian	[aː]	father
[ei]	name	[æ]	back

For the graphemes *ae, air, ai, au, ay,* etc., there is a vast number of different pronunciations depending on left and right contexts. There are general rules that can be used for many words, and specific ones which can be applied only for a few words. Pronunciations of words like: *cycle* [saikl] versus *bicycle* ['baisikl] or *children* ['tʃildrən] versus *childish* ['tʃaildiʃ] are difficult to understand when learning the language.

Another well-known problem for English is the need to take into account the morpheme boundaries of words. *Boathouse* has to be recognized as *boat* and *house*; otherwise *th* would be transcribed as *th* in *marathon*, and the *th* of **cathouse** as the *th* in **catholic**. There are many compound words like *knothole, pothole, hothouse, houseboat, handmade, cheesecake, fireplace, driveway, earthrise,* and *leadership.*

Knowing the primary stressed syllable of a word is a basic necessity if an English word is to be pronounced correctly. This is a real difficulty for non-native English speakers. This is particularly true for French-speaking people who tend not to reduce the unstressed syllables when speaking English. For instance, in the following words where the stressed syllables are in bold type: **pho**tograph, pho**to**grapher, photo**gra**phic, or e**co**nomy, eco**no**mic, it is very important to know the primary stressed syllable not only to pronounce the stressed syllables correctly, but also the unstressed syllables, sometimes reduced to a schwa.

photograph	'fəʊtəgræf	economy	i'kɒnəmi
photographer	fə'tɒgrəfə	economic	iːkə'nɒmik
photographic	fəʊtə'græfik		

Determining the primary stressed syllable of a word is not an easy task. There are some rules, but in some cases the stressed syllable of a word has to be learned, as for instance in ba**na**na versus **Ca**nada. Some suffixes can help determine the stressed syllable: words ending in *ic* have a primary stressed syllable on the syllable preceding *ic* as in photo**gra**phic and eco**no**mic. Depending on the grammatical category of the word, the stressed syllable may vary as shown in the examples below:

	verb	**noun or adjective**
use	[juːz]	[juːs]
record	[ri'kɔːd]	['rekəd]
elaborate	[ilæbə'reit]	[i'læbərit]]

The stressed syllable can also vary by geographical area. According to the American Heritage dictionary, the pronunciation of *banal*, for example, is not settled among educated speakers of American English, and several pronunciations are widely used.

For borrowed words from another language, the pronunciation is very often not the original one: sometimes phonemes of the original language do not exist in English and have to be replaced by the nearest phonemes; *bonjour* has a nasal /õ/ phoneme unknown in English and replaced by /ɒn/. English will tend to stress the first syllable, so the English *bonjour* is very different from the French *bonjour*. The word is partially assimilated. Other words from French origin have kept their final stressed syllables as in *canal, Pascal, entrée, bourgeois*, etc. These words are pronounced neither according to the rules of standard English nor the rules of the original French language.

The pronunciation of an isolated word may be ambiguous, and depends on the semantic context.

read	/riːd/	*read*	/red/
live	/liv/	live	/laiv/
bass	/beis/	*bass*	/bæs/

Attempts have been made to solve the ambiguities by examining the local context (e.g., bass fishing, bass guitar, I have read, to read, I live, five lives), or a larger context (Yarowsky, 1994).

3.2   French

In French, the [ʃ] phoneme can result from the conversion of:

ch	*chat*
sch	*schéma, putsch*
sc	*fasciste*
sh	*flash, shampooing, short*

The grapheme string *ch* can lead to the following phonemes:

[ʃ]	*chat, machine, architecte*
[k]	*bronchoscopie, chlore*

So, to a certain extent as in English, there is a large gap between spelling and pronunciation. Many words are homophones. The following words have the same pronunciation:

[sɛ̃]   *sain, ceint, saint, ceint, seing*
[sɛ]   *ces, ses, sait, c'est, saie*
[vɛr]   *vair, ver, verre, vers, vert*

This is a source of difficulties for young people learning how to write the language. It is also the source of many plays on words (*le ver vert va vers le verre vert*).

In some cases, a word can be made up with two juxtaposed words each of these keeping its pronunciation as in *télésiège, entresol*. Usually *s* between two vowels is pronounced [z] as in *Asie*, but not in these words that are in fact the concatenation of a prefix and a word: *télé+siège, entre+sol*. In French, this happens mostly with prefixes and words.

French is not a stressed language. The final syllable of words is stressed, but the other syllables are not really reduced as in English. The duration of unstressed syllables is reduced in polysyllabic words, but the phonemes remain unchanged as for instance in *papa, bonbon*, or *Tintin*.

In French, in some contexts, a phoneme is added between two words that are pronounced together in the same breath group or syntagm as in *nous avons, les avions, chez elle, un enfant*. This could occur when the last consonant of the previous word's ending is *n, s, d, t, x* or *z*, and the first syllable of the next word is a vowel or an aspirated *h*. It is called a liaison. Depending on the context, and mostly on the grammatical categories of both words involved, this liaison can be mandatory, optional or illicit.

In contrast, a mute *e* is usually elided at the end of a word (*belle*) or in the middle of a word if eliding the *e* results in a word that is still pronounceable (*retard, boulangerie, tellement*). This elision can be optional if the speech rate is very slow.

As in English, the same word can have different pronunciations depending on the meaning: *as* ([a] vs. [as] in *tu as; un as*), *relations* ([ʀ(ə)lasjɔ̃] vs. [ʀ(ə)latjɔ̃] in *les relations; nous relations*), *président* ([pʀezidɑ̃] vs. [pʀezid] in *le président; ils président*). There are many examples in French. Often, local context can resolve the ambiguity. A definite solution to that problem would be to do a parsing of the sentence considering the grammatical category of each word in the sentence. This parsing is complicated, so usually not done.

Borrowed unassimilated words are also a problem in French. Resulting from the massive borrowing from English of words ending in *ing*, a new [ŋ] phoneme is used by most French-speaking people. Proper names in French include a large percentage of names that are correctly transcribed by the set of letter-to-sound rules for words (Bouvier, Divay, Martin). Nevertheless, some French-sounding proper names are the concatenation of two words (Lesage, Montmartre, Bourgneuf, Després), each of these retaining its own pronunciation. The problem is similar to compound words with prefixes mentioned above. For foreign unassimilated or partially assimilated proper names, the problem is still open. Attempts have been made to recognize the original language of a proper name, and then to apply a set of rules depending on that language. The real pronunciation is often different from the original pronunciation, especially if some of the phonemes do not exist in the target language. Proper names like *Thatcher* or *Smith* are very often pronounced with a phoneme [s] for *th* as French does not have a [θ] phoneme.

3.3 Remarks For Both English And French

Spelling and pronunciation are for both languages the result of history. The pronunciations continue to evolve, but dictionaries and conservatism have frozen the spelling, resulting in a gap between what is spoken and what is written. Several spelling reforms have been suggested for both English and French, but until now, without any success. An Initial Teaching Alphabet has even been designed to help young children read English.

Text-to-speech synthesizers must be able to synthesize any input text. They usually have a first step of text normalization, replacing grapheme characters by other grapheme characters or phonemes. The normalization step has to replace logographs by their equivalent graphemes (*24* replaced by *twenty-four* for English or *vingt-quatre* for French; & replaced by *and* for English or *et* for French), and to pronounce or spell abbreviations (km, kg, Dr.) and acronyms (IBM, USA). In some cases, the user may want to have the punctuation spelled. This is a preprocessing step of the letter-to-sound module.

Another difficulty concerns compound words that should be considered and pronounced as one word in spite of the space between the words. This is true for compound words like *coffee cup* in English, or idioms like *de temps en temps, pomme de terre,* and *tout à coup* for French.

4. RULE FORMALISM

An expert system is a piece of software that tries to infer new information from initial facts by applying a set of rules specific to one domain. The system is designed to apply rules, knowing only the formalism of the rules. The expert uses this formalism to write a set of rules representing the knowledge of the expert system. The human being is an expert in different domains; one of them is reading aloud a text written in his or her native language. By experience, he or she uses rules to convert text into sounds, or to recognize words globally, such as *monsieur* [məsjø] in French, where the spelling is almost logographic.

The rewriting rule formalism is well adapted to letter-to-sound conversion. Input text can be replaced in different steps by other graphemes, and then by phonemes, and even phonemes can be replaced by other phonemes to take into account rules more appropriate at the phoneme level, or speech rate, for instance. The general synopsis shown in Figure 1 consists of rules written according to the syntax of the formalism. These rules are compiled into an internal format. The interpreter uses these compiled rules to transcribe a grapheme text into a phoneme string.

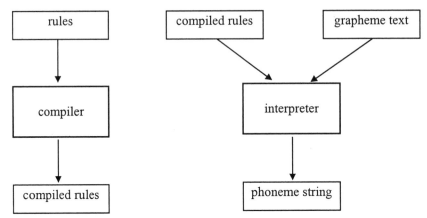

Figure 1.   General synopsis of a rule-based expert system.

4.1   External Code

The legal graphemes and phonemes of the language are then listed and an external code is defined for each unit. The graphemes are the characters you would find in a regular text (alphabetic characters, digits, punctuation). The phonemes are the different sounds the synthesizer will have to produce. Making an inventory of the different phonemes is not that straightforward. There are many differences between the various dictionaries of English for both the number and the external code of phonemes. The main differences are often for reduced vowels, syllabic *n* or *l*, or pronunciation of *r*. See the phoneme list for French and English in the Appendix. This external code can easily be redefined. It should be as obvious as possible for humans.

To explain the formalism irrespective of any language, basic input and output units are defined as follows:

$S = A, B, C, D, E, F, G, H/$;   input and output characters

where each letter represents one grapheme or one phoneme.

4.2   Strings and Classes

The use of classes is mentioned later. It is a set of strings having a common property. A string is the concatenation of graphemes or phonemes; for instance, ABA is a string made up of elements of S.

Suppose:
    'Cl1', 'Cl2' and 'Cl3' are classes formed with strings made up of elements of S. They could be defined, for instance, as:

    'Cl1' : F,   EA,   G/

'Cl2' : C,   BA,   FFB/
'Cl3' : BAB,   DEF/

'Cl1' is a shortcut representing the string F, EA or G.

4.3   Blocks of rules

The entire set of rules can be divided into different steps, each step corresponding to a block of rules processing the input text from either left-to-right or right-to-left. Each block of rules has a number and a tag indicating if the scan is left-to-right (default), or right-to-left.
Example:
    begin {Block *i*}
            rule 1
            ......
            rule *n*
    end

4.4   Rules

A rewriting rule consists of:
    <label>: <ls>   → *<rs>*   / *<lc>*   - *<rc>* ;
    where:
            <label>   is the rule number,
            <ls> (left string)   is the string to be replaced,
            <rs> (right string)   is the string replacing *ls*,
            <lc> (left context)   represents the strings to be found on the left side of *ls*,
            <r>c (right context)   represents the strings to be found on the right of *ls*.

<lc> and *<rc>* are formed with operands (characters, strings, classes) and operators (concatenation, logical or, negation).

Example :
    10 : f   → *v*   / *_o*   - *_* ;
    f is replaced by *v* if *f* is preceded by *_o* and followed by *_*, where _ means a space. This rule would replace *f* by v in *_of_* resulting in *_ov_*.

4.5   Interpreting the Rules

4.5.1   Buffers

If the left and right contexts of a rule match an *ls* string of the input text, the *rs* string replaces the *ls* string. This replacement can be done by replacing the input *ls* string, or by copying the *rs* string to an output buffer, keeping the input text intact. If the scan is left to right, with one buffer, the left context has to be written according to the rules having processed this left context. With two buffers, the input text is not altered during the processing of the block, and in that case, the left context is written regardless of the rules applied to the left of *ls*. If the scan is left-to-right, the left context of the output buffer (the right context in a right-to-left scan) can also be tested.

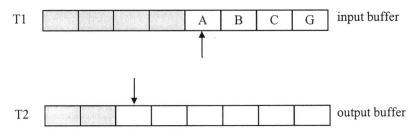

Figure 2.   Input and output buffers.

4.5.2   Processing the rules

The rule having the longest match between the set of all the *ls* strings of the block, and the string beginning with the next character to be processed in the input text, is searched for first. If both contexts are true, the rule applies; otherwise, another rule is searched for, first any other rule with the same *ls*, and then in decreasing length of *ls* matches.

Let us consider the following rules:

```
begin
  150: AB          → ... ;
  151: A           → ... ;
  152: ABC         → ... ;
  153: AB          → ... ;
  154: BC          → ... ;
  155: ABCH        → ... ;
end
```

and the input string, *ABCG*, to be processed (see Figure 2).

The longest match between the left strings (*ls*) of the rules in the block, and the input string to be processed is searched. In this case the longest match is *ABC*. So, rule 152 is tested. If the contexts are true, the rule is applied, and the next character to be processed is *G* in the input buffer. If the context is false, the rules are tested in decreasing order of the longest match. Rules with *AB* as *ls* are tested in the order in which they are written (150, 153). Then if no rule has yet been applied, rule 151 is tested. If no rule is true, the first character *A* to be processed is copied into the output buffer, and the procedure starts again with the next character *B*. The order in which rules are tried is: 152, 150, 153, 151. The order in which the rules are written is significant only for those having the same *ls*.

Using this expert system formalism, the expert has to develop a set of rules to transfer his or her knowledge to the system.

4.6   Formal Examples of Rules

Classes 'Cl1', 'Cl2', 'Cl3' are previously defined in paragraph Strings and Classes.

   1: AB   → G   / E.'Cl1'   - ;

   AB is replaced by G if, on the left side of AB, an element of the class 'Cl1' is
found preceded by E. If the input string to be processed is: EEA**AB,** the rule applies
for **AB** which is replaced by **G** in the output buffer.

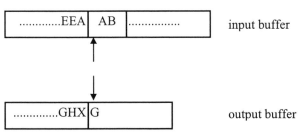

Figure 3.   Left context for rule 1.

   2: CB, EF   →   / 'Cl3'.H   - 'Cl2'.G, 'Cl3' ;

   The string **CB** or **EF** is deleted (replaced by an empty *rc*) in the following
contexts:
- on the left of CB or EF, an H preceded by an element of the class 'Cl3', and
- on the right of CB or EF,
      either an element of 'Cl2' followed by a G,
      or an element of 'Cl3'.

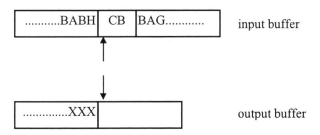

Figure 4.   Left and right contexts for rule 2.

   3: EE   → G   / Non (A, B)   - ;
   EE is replaced by G if the left context is not an A, or a B.

4: AD   → E   / <C.'Cl1'> B   - ;

AD is replaced by E if:

- the left context of the output buffer (between angle brackets) is an element of 'Cl1' preceded by C, and
- the left context of the input buffer is B.

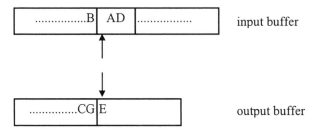

input buffer

output buffer

Figure 5.   Left context in the output buffer.

5.   EXAMPLES OF RULES FOR ENGLISH

A few examples of the rules are given for English to illustrate the way problems previously mentioned are coped with.

Example 1: words beginning with *ba*

A few words beginning with *ba* have an *a* that is pronounced [ei] as in *baby, bacon, bagel, basal, basic,* and *basis.* Other words are pronounced using [æ] (*Bali, Babylon*) or another phoneme depending on the right context as in *bald, bar,* and *ball.*

A class of word beginning with *ba* with *a* pronounced [ei] can be defined as:

'Bei' : bel, by_, con, gel, sal_, sic, sin, sis /   _ means a space

and the rule can be written:

5: a   → [ei]   / _b   - 'Bei' ;

The rule applies for *Baconian* and *basal,* but not for *Babylon* and *basaltic. Babying, babyish, babyhood* would have to be broken up into *baby* and a suffix *ing, ish* or *hood.* Classes could be defined in a similar way for words beginning in *da, fa,* etc., to take into account words like *cajun, danger, fatal, gable, haven, jacob, label, mania, nasal, quasar, racial, salient, talus* and *vacancy* for instance.

Example 2: words ending in *ace*

Words ending in *ace* often have an *a* which is pronounced [ei] as in *face, space, carapace,* and *embrace.* This is not the case for *menace, furnace, surface, palace, preface,* and *necklace* where the first syllable is stressed, so *a* is reduced to [i]. A class of words ending with *ace* with *a* pronounced [i] can be defined as:

'P_ace' : pref, surf, resurf, pal, neckl, sol, popul, men, pinn, furn, hor, terr/

This class is used by the following rule:
   6: a   → [i]   / _.'P_ace'   - ce_ ;
   & surface, palace &

Example 3: words ending in *age*

A similar rule could be defined for words ending in *age*. Usually in these words, the *a* is an unstressed [i] as in *advantage, assemblage*, and *bandage*, but not in *engage* and *outrage* where *a* is pronounced [ei], or in *barrage* and *entourage*, where *a* gives [aː].
Rules could be written as follows:

'P1_age' : eng, greeng, outr, offst, ramp, teen, enr, upst/
'P2_age' : barr, camoufl, coll, cors, décollet, espion, entour, fusel,
       mén, mir, mass, mont, persifl, pl, pot, sabot, report/
'C' : b, c, d, f, g, h, j, k, l, m, n, p, q, r, s, t, v, w, x, z /   grapheme consonants
   7: a   → [aː]   / _.'P2_age'   - ge_ ;
   & reportage, barrage &
   8: a   → [ei]   / _ , _.'C', _.'C'.'C' - ge_ ;
   & cage, stage &
   9: a   → [ei]   / _.'P1_age'   - ge_ ;
   & teenage, engage &
   10: a   → [i]   /   - ge_ ;
   & advantage, assemblage &

Some of the words of 'P1_age' could be broken up into morphemes (out+rage, off+stage) previously, and consequently deleted from the class 'P1_age'.

Example 4: general rules

In most cases, an *a* followed by a consonant except an *r*, and an *e* at the end of a word is pronounced [ei] as in: *babe, cake, space, cascade*, and *chicane*. Rules of Examples 2 and 3 are exceptions to that rule. A more general rule dealing with words ending in: a.'C'.e_ can be written as follows:
   'C-r' : b, c, d, f, g, h, j, k, l, m, n, p, q, s, t, v, w, x, z / consonants except r
   11: a   → [ei]   /   - 'C-r'.e_ ;
a is rewritten [ei] if followed by a *consonant except r*, and an *e* followed by a space. The rule does not apply for *bare, care*, or *square*.

Other more or less general rules, and examples are given below.
   12 : aa   → [aː] ;
   & aachen, afrikaans, baa, bazaar &
   13 : ae   → [ɛə]   /   - r.'V' ;
   & 'V' means a vowel; aerate, aerial, aeronaut &
   14 : ae   → [iː]   / _   - ;

& aegis, Aesop, aesthete &
15: ae   → [ei] ;
& Israel, sundae, reggae &
16: ai   → [ei][i]   /   - c_ ;
& archaic, judaic, algebraic; 38 words ending in aic in the American Heritage
(AH) CD-ROM &
17: ai   → [ɛə]   /   - r ;
& airing, airless, affair, stair &
18: au   → [aː]   / dr, l   - ght ;
& draught, laughter &
19: au   → [oː] ;
& audio, baulk, daughter, faun &
20: ay   → [ei] ;
& away, day, say, stray; 776 words ending in ay in the AH CD-ROM &

21: cc   → [k][s]   / a, e, o, u   -   e, i ;
& accident, eccentric, occident, success &
22: cc   → [k] ;
& acclaim, accolade, accuse &

For consonants, the list of exception rules is relatively short, but for vowels, it can
get very large.

Example 5: *morph decomposition*

   Words ending in *ed* are often the past tense of a verb, like *edited, aborted,*
accused, classified, and *accepted,* but not always as in *bed, abed, biped,* and *moped.*
The following rule:
   23: ed   → _+_[ə][d]   / ss, pt, k   - _ ;
   decomposes *accessed* into *access_+_[ə][d]*
   Rule 23 works for *dressed, regressed, tossed, accepted, adopted,* and *walked*
Rule 24:
   24: ed   → e_+_[ə][d]   / c, s, ag, dg, bl   - _ ;
   breaks up *accused* into *accuse_+_[ə][d]*
   Rule 24 works for *placed, damaged, judged,* and *doubled.*
In fact, left contexts for rules 23 and 24 are classes with all the left context strings
the rule applies for.

The same classes can be used to break up words ending in *ing,* like *accusing* which
decomposes into *accuse_+_[ɪ][ŋ],* ment *(alignment, achievement),* ful *(careful,
doubtful),* and *hood (childhood, neighborhood),* etc.

Example 6: *stressed and reduced vowels*

Vowels can depend on the stressed syllable. *papa* is transcribed as follows: the second *a* is replaced by *a* plus a mark *[L1]* indicating that the stress is on the vowel on the left of the *a* at the end of the word. Both *a* are then transcribed into the pseudo-phoneme [æ|ə] resulting in [p][æ|ə][p][æ|ə]. [æ|ə] means it is either an [æ] or an [ə]. This pseudo-phoneme is declared in the code section (See paragraph External Code). Then the stressed syllable is replaced by [æ], and the unstressed syllable by [ə] (['pæpə]).

25: a   → [L1]a   / 'C'   - _;
26: a   → [æ|ə] ;

Example 7: *morph decomposition and stress: effect of suffixation on stress*

Words ending in *ied* are very often the past tense of a verb: *classified, notified, citied, cried, fried,* and *curried*; 351 words end in **ied* in the AH CD-ROM.

27: ied   → y_+_[d]   /   - _;
so *classified* is rewritten as *classify_+_[d]*.
28: y   → [L2]y   / f   - _;
meaning *y* is rewritten *y* but with a L2 stress mark indicating that the primary stress is on the syllable two syllables on the left of the *y* syllable, e.g. on *class* of *classify*.

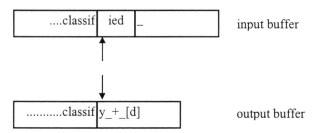

Figure 6.   Morph decomposition for classified.

Then the words are converted by grapheme-to-phoneme rules resulting in [k][l][æ|ə][s][i|ə][f][ai]_+_[d]; then in [klasefaid] when the stress is taken into account. The mark _+_ is deleted by rule.

Others examples:

29: i   → [L1]i   /   - c_ ;
economic: stress is one syllable left of *ic*
30: o   → [L1]o   /   - my_ ;
economy: stress is one syllable left of *omy*

So, some suffixes can impose the stressed syllable as mentioned in the following examples.

Stress on the last syllable:

ee	: refugee, employee
eer	: career, mountaineer
ese	: Chinese, Japanese
oo	: bamboo, taboo
oon	: baboon, balloon
ette	: cigarette, kitchenette
esque	: grotesque, romanesque
ade	: grenade, cascade

Stress 1 syllable on the left (L1):

ic	: phonetic, republic	ious	: industrious
ical	:grammatical, economical	eous	: courageous
ically	: economically	ual	: intellectual
ion	: attention, aggravation	ity	: ability, stupidity
ional	: professional	logy	: biology, apology
ionally	: professionally	logist	: biologist
ionary	: revolutionary	graphy	: geography
ian	: musician, vegetarian	grapher	: photographer
ial	: provincial		

Stress 2 syllables on the left (L2):

ous   : autonomous, miraculous, amorous

Example 8: from phonemes to phonemes

In some cases, phonemes can be replaced by other phonemes in a final block of rules:

31: [e][d]   → [t]   / [f]_+_, [k]_+_, [p]_+_, [s]_+_ - _ ;

walked already transcribed into *[w][ɔː][k]_+_[e][d]* becomes the phoneme string [wɔːkt] after removing the morpheme mark.

Example 9: prefixes

Prefixes are also stripped from the root. *multimedia* would be broken up as mentioned below:

32: multi   → [m][ʌ][l][t][i]_+_   / _ - ;

multimedia becomes [m][ʌ][l][t][i]_+_media.

Then media is converted by rules. The stress is on the left of *ia* for words ending in *ia (California, sinfonia, minutia)*. The morpheme boundary is deleted: [mʌltiˈmiːdiə]. For multi*e*thnic, it is important to recognize the decomposition multi+ethnic, and not to apply rules to process *ie* as in p*ie*ce.

6. EXAMPLES OF RULES FOR FRENCH

A few examples of rules are given for French to deal with problems previously mentioned. The formalism stays the same. A new rule set has to be designed.

Example 1: vowels

a	is pronounced	[a]	in *papa, la, table*
ai	is pronounced	[ə]	in *bienfaisant, contrefaisait, défaisait, faisan, satisfaisant*
			but not in *faisceau, chauffais* where it is an [ε]
ai	is pronounced	[ε]	in *lait, était, domaine, capitaine, lainage*
ain	is pronounced	[ɛ̃]	in *sain, pain, crainte, vaincre, maintenir, américain, ainsi*
			but not in *saine* where the rule for *ai* applies.

The rules could be written as shown below.
'C' : b, c, d, f, g, h, j, k, l, m, n, p, q, r, s, t, v, w, x, z / grapheme consonants
'V' : a, à, â, æ, e, é, è, ê, ë, i, î, ï, o, ô, ö, u, ù, û, ü, y / grapheme vowels
33: ain   → [ɛ̃]   /   - 'C', _ ;
ain gives [ɛ̃] if *ain* is followed by an element of 'C' or a space.
34: ai   → [ə]   / f   - s.'V' ;
ai is pronounced [ə] if *ai* is preceded by *f* and followed by *s*
and an element of the class 'V'.
36: ai   → [ε] ;
37: a   → [a] ;

Example 2: consonants

In French, *c* is pronounced [s] if followed *by e, i, é, è, ê, î* as in *centre, cigarette,* and *accès*; otherwise it is pronounced [k] as in *case* and *cobra*. *g* gives [ʒ] before the same vowels (*agent, agile*), otherwise [g] (*algorithme, garage*). The rules are the following:
'VaCG' : e, i, é, è, ê, î / Vowels after C or G
38: c   → [s]   /   - 'VaCG' ;
39: g   → [ʒ]   /   - 'VaCG' ;

Example 3: morpheme boundary

Usually, as a general rule, *s* is pronounced [s] as in *saint, salon, sérum, sofa, absent,* and *seconde*. There are exceptions to this general rule. *s* between two vowels is pronounced [z] like in *basalt, vase, maison,* and *anglaise*. But there are exceptions to these exceptions. If *s* follows a prefix, it produces an [s] as in *télésiège, antisocial, autosuggestion, microseconde,* and *parasol*. The rules could be written as shown below.

'Prefix' : anti, auto, micro, para /   Class of prefixes
40: s   → [s]   / _.'Prefix'   - 'V' ;
41: s   → [z]   / 'V'   - 'V' ;
42: s   → [s] ;

Example 4: ends of words

d at the end of words is usually not pronounced as in *laid, pied, nord, quand, renard, nid,* and *blond.* In Le Petit Robert CD-ROM dictionary (Rey et al., 1989), 527 words end in *d*; 279 words end in *ard,* with a *d* not pronounced. *d* is pronounced in a few words mostly of foreign origin, such as *bagad, djihad, farad, lad, rad, skinhead, tansad, baroud, barmaid, bled, fjord, lord, raid, stand,* and *sud.* The rules could be formulated as:
   'D' : barou, barmai, ble, fjor, lor, rai, stan, su/
   43: d   → [d]   / a   - _ ;
   d at the end of a word is pronounced [d] if *d* is preceded by *a.*
   44: d   → [d]   / _ . 'D'   - _ ;
   d is pronounced [d] if *d* if preceded by an element of 'D'
   preceded by a space, and followed by a space (end of word).
   45: d   →   /   - _ ;
   Otherwise *d* is not pronounced at the end of a word.
Rules 43 and 44 could be merged into one rule with the left context: a, _.'D'.

Example 5: beginnings of words

   46: b   → [p]   / _a, _o, _su   - s, t ;
means that *b* preceded by *a, o,* or *su* at the beginning of a word, and followed by *s* or *t,* is replaced by the phoneme [p] as in *abscisse, absence, obscur, obtenir,* and *substituer.* Considering the derived forms, the Petit Robert dictionary has 271 words beginning in *abs,* 295 in *obs,* 209 in *subs,* 0 in *abt,* 128 in *obt,* and 54 in *subt.* This is a general rule but with a few exceptions like *subsister* or *subtropical.*

Example 6: geminates (twice the same consonant as in dd, tt)

   In order to eliminate geminates, one possibility is to analyze the last character sent to the output buffer.
   47: d   →   / < [d] >   - ;
d is eliminated if the left context in the output buffer is already a [d] phoneme.

Example 7: text normalization: from graphemes to graphemes

   Text normalization consists of replacing numbers, abbreviations, and acronyms by their full text equivalents. Both input and output are graphemes. Normalization is handled by an optional preprocessing block of rules.

Numbers. 157 is rewritten as *cent cinquante sept* by a set of rules checking the left and right context for each digit.

'Digit' : 0,1, 2, 3, ..., 9/ is the class for digits

48: 1	→ cent_	/	- 'Digit'.'Digit'._ ;
49: 5	→ cinquante_	/	- 'Digit'._ ;
50: 7	→ sept	/	- _ ;

1 is rewritten *cent_* if followed (right context) by two digits and a space, etc.

Abbreviations. kg for *kilo, Dr* for *Docteur, Pr* for *Professeur, bd* for *Boulevard*, etc.

| 51: kg | → kilos | / _, 'Digit' | - _ ; |

kg is replaced by *kilos* in *5kg* or *trois kg*.

Acronyms. Similar rules are used to spell the acronyms (I.B.M. gives [ibeɛm]):

| 52: B. | → bé | / _, . | - ; |

B followed by a point is replaced by *bé* (spelled) if B is preceded by another point or a space. In I.B.M., or *vitamine B.*, *B.* is correctly spelled.

For acronyms without the points (IBM for instance), a special procedure checks the characters and decides if the word is pronounceable, or must be spelled. It is not done by rules.

Preprocessing procedures not based on rules are also used in cases like $50 which gives: *cinquante dollars* and where $ and 50 have to be permuted.

Example 8: *liaisons*

As explained previously, a new phoneme is added between two words in some cases. Without knowing the grammatical categories of the words, the system has to rely on classes listing the words frequently involved in that liaison problem. Examples:

'Li_Z' : les, des, nous, vous, ils, elles, ..., chez, ... /
'Li_N' : mon, ton, son, un, en, .../
'LI_T' : petit, grand, second, dont, .../

53: _	→ _[z]	/ _.'Li_Z'	- 'V' ;
54: _	→ _[n]	/ _.'Li_N'	- 'V' ;
55: _	→ _[t]	/ _.'Li_T'	- 'V' ;

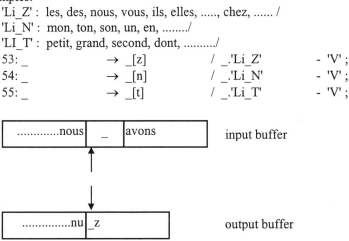

Figure 7. Liaison for French.

The space between *nous* and *avons* gives _[z] in the output buffer.
Rule 53 is applied for *nous avons, les avions,* and *chez elle.*
Rule 54 is applied for *un enfant* and *mon avion.*
Rule 55 is applied for *petit avion* and *quand il.*

Example 9: *elision: from phonemes to phonemes*

Some rules, mostly rules dealing with mute *e* and semivowels, can be more easily expressed on the phoneme strings. A block of rules is run after the grapheme-phoneme conversion.

'VP' : [a], [i], ..., [ε], ... /   the vowel phonemes
'CP' : [b], [d], ..., [l], ... /   the consonant phonemes

56: [ə]   →   / 'VP'.'CP'   - ;
[ə] is elided if preceded by a consonant phoneme,
preceded itself by a vowel phoneme.

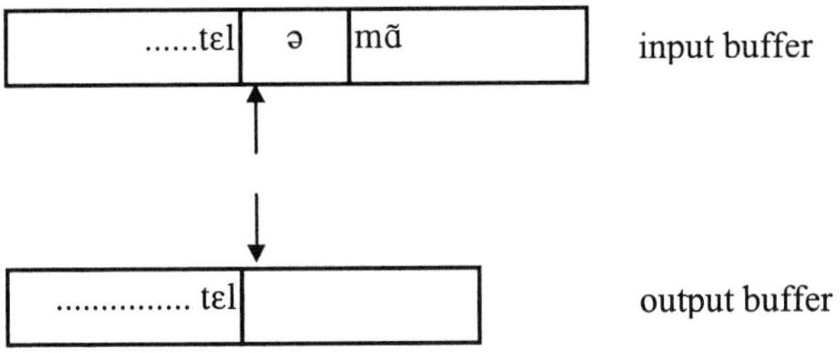

Figure 8. Elision for French.

Mute *e* is also eliminated before a vowel phoneme and after a consonant phoneme followed by a vowel phoneme as in *emploiera* ([ãplwaəra], which becomes [ãplwara]). Elision often occurs at the end of words (*belle*), or in the middle of words (*emploiera, tellement*). It can occur in the first syllable (*pesanteur, retard, teneur*) except when there are two consonants as in *premier*. It never occurs if suppressing [ə] would result in three or more consecutive consonants. It never occurs for some words, such as *femelle, benoit,* and *guenon.*

7.   CONCLUSIONS

The difficulties of grapheme-to-phoneme conversion for English and French are a reality for young people learning their native language, and for foreigners learning it as a second language. The two languages have evolved from different origins, and their pronunciations have been considerably modified over the centuries. Spelling has evolved more slowly, creating a gap between what is written and what is pronounced. Both languages have interacted with each other and continue to do so.

Developing speech synthesis software for English or French requires teaching the computer all the details and idiosyncrasies of each language (for English: Bernstein and Nessly, 1981; Hertz, 1985; Klatt, 1987; O'Malley, 1990; for French: O'Shaughnessy, Lennig, Mermelstein, and Divay, 1981; Divay, 1984, 1990; Divay and Vitale, 1997; Boula de Mareüil, 1997). Rules are written in an ad hoc formalism understandable by the expert developing the set of rules, and easily usable by the computer. Each language has a large set of rules for converting the different words into phonemes. French presents some specific difficulties with liaisons and elisions. Morpheme decomposition and stressed syllables complicate the conversion for English. For both languages, the non-homophone pronunciation of homographs is difficult to solve without a complete parsing of the sentence.

Words not correctly transcribed by rules are inserted into a dictionary of exceptions. The dictionary can also be used to get more information on words: grammatical categories, morphs and roots, and stressed syllables (Laporte, 1988). MIT (Allen et al., 1979) used a 12,000 morph lexicon to do the morph decomposition. In English, high-frequency words are sometimes exceptions to the general rules (two, for, of). As memory has become less expensive, speech synthesizers have tended to incorporate bigger (exception) dictionaries.

Developing a set of rules is time-consuming, so attempts have been made to use learning algorithms to explicitly or implicitly discover the relationship between graphemes and phonemes (Golding, 1991; Meng, 1995; Yvon, 1996). Although promising, they have not (yet) reached the level of rule sets developed by humans. The automatic determination of the underlying structure of a language is not easy, and nor is the development of a universal rewriting rule formalism for the various languages.

REFERENCES

Allen, J., Carlson, R., Granström, B., Hunnicutt, S., Klatt, D. H., and Pisoni, D. B. (1979). *Conversion of unrestricted text-to-speech* (Unpublished Monograph). Cambridge, MA: Massachusetts Institute of Technology.

American heritage talking dictionary (3rd ed.) [CD-ROM] (1994). Softkey, Cambridge, MA.

Ben Crane, L., Yeager, E., and Whitman, R. (1981).   *An Introduction To Linguistics*. Boston: Little, Brown and Company.

Bernstein, J., and Nessly, L. (1981). Performance comparison of component algorithms for the phonemicization of orthography. *Proceedings of the 19th Annual Meeting of the Stanford University Association for Computational Linguistics*. Stanford, CA: Stanford University.

Boula de Mareüil, P. (1997). *Etude linguistique appliquée à la synthèse de la parole à partir de texte*. Thèse, Université Paris XI, Orsay, France.

Catach, N. (1978). *L'orthographe*. Collections "Que sais-je?" Paris: Presses Universitaires de France.

Encarta: The Complete Interactive Multimedia Encyclopedia [CD-ROM] (1995). Redmond, WA: Microsoft.

Divay, M. (1984). *De l'écrit vers l'oral ou contribution à l'étude des traitements des textes écrits en vue de leur prononciation sur synthétiseur de parole.* Thèse d'Etat, Université de Rennes, Rennes, France.

Divay, M. (1990, March). Traitement du langage naturel: la phonétisation ou comment apprendre à l'ordinateur à lire un texte Français. *Micro-Systèmes*, 139-143.

Divay, M., and Vitale, A. J. (1997). Algorithms for grapheme-phoneme translation for English and French: Applications for database searches and speech synthesis. *Computational Linguistics, 23*(4), 495-523.

Golding, A. R. (1991). *Pronouncing names by a combination of case-based and rule-based reasoning.* Ph.D. Thesis, Stanford University, Stanford, CA.

Hakoda, K., Kitai, M., and Sagayama, S. (1997). Speech recognition and synthesis technology development at NTT for telecommunications services. *International Journal of Speech Technology, 2*(2), 145-153.

Hertz, S. R. (1985). A versatile dictionary for speech synthesis by rule. *Journal of the Acoustical Society of America 77* (Suppl. 1), S11.

Hunnicut, S. (1976). Phonological rules for a text-to-speech system. *American Journal of Computational Linguistics*, Microfiche *57.*

Klatt, D.H. (1987). Review of text to speech conversion for English. *Journal of the Acoustical Society of America, 82*(3), 737-793.

Laporte, E. (1988, May). *Méthodes algorithmiques et lexicales de phonétisation de textes.* Thèse, Université Paris 7, Paris, France.

Meng, H. M. (1995). *Phonological parsing for bi-directional letter-to-sound and sound-to-letter generation.* Unpublished doctoral dissertation, Massachusetts Institute of Technology, Cambridge, MA.

O'Malley, M. H. (1990, August). Text-to-speech conversion technology. *IEEE Computer*, 17.

O'Shaughnessy, D., Lennig, M., Mermelstein, P., and Divay, M. (1981). Simulation d'un lecteur automatique du Français. *12 èmes Journées d'Études sur la Parole* (pp. 315.325). Montreal, Canada.

Rey, A., Duval, A., Vienne, B., Struyf, B., Divay, M., Lootens, T., and Zimmermann, S. (1989). *Le Robert electronique, ensemble d'outils d'aide à la rédaction de textes Français sur disque optique compact* [CD-ROM], Paris: Dictionnaires Le Robert..

Taurus (1993). *Système de gestion electronique de documents multimédia. Guide d'évaluation.* Pantin, France: DCI (Dron Concept Informatique).

Wells, J.C. (1982). Accents Of English: An Introduction. Cambridge, MA: Cambridge University Press.

Yarowsky, D. (1994). *Homograph disambiguation in text-to-speech synthesis.* Proceedings of the Second ESCA/IEEE Workshop on Speech Synthesis (pp.157-172).

Yvon, F. (1996). *Prononcer par analogie: motivation, formalisation et évaluation.* Thèse, Ecole nationale des Télécommunications, Paris.

Appendix: Phonemes for French and English

French consonants

[b]	Bon
[d]	Dans
[f]	Fin
[g]	Gare
[ʒ]	Je
[k]	Kilo
[l]	Lent
[m]	Main
[n]	Nous
[ɲ]	aGNeau
[ŋ]	campiNG
[p]	souPe
[r]	Rat
[s]	Sale
[t]	Terre
[v]	Vous
[z]	Zéro
[ʃ]	CHat

French vowels

[ɑ]	pâte
[a]	plAt
[ə]	lE
[e]	blé
[ɛ]	père
[i]	Il
[ø]	pEU
[œ]	cœur
[o]	tAUpe
[ɔ]	bOl
[y]	rUe
[u]	rOUe
[w]	OUi
[ɥ]	hUile
[j]	Yaourt
[ã]	sANs
[õ]	bON
[ɛ̃]	matIN
[œ̃]	brUN

English consonants

[p]	pop
[t]	tight, stopped
[k]	kick, cat,
[b]	bib
[d]	deed, milled
[g]	gag
[f]	fife, phase,
[s]	sauce
[ʃ]	ship, dish
[v]	valve
[z]	zebra, xylem
[ʒ]	vision, pleasure
[m]	mum
[n]	no, sudden
[ŋ]	thing
[l]	lid, needle
[r]	roar
[tʃ]	church
[dʒ]	judge
[h]	hat
[θ]	thin
[ð]	this

English vowels

[w]	with
[j]	yes
[iː]	bee
[i]	pit
[e]	pet
[aː]	father
[æ]	pat
[ɔː]	caught, paw, for
[ɒ]	pot, horrid
[ʊ]	took
[uː]	boot
[ə]	about, item,
[ʌ]	cut
[ei]	pay
[ɛə]	care
[ai]	pie, by
[iə]	pier
[əʊ]	toe, hoarse
[ɔi]	noise
[aʊ]	out
[ɜː]	urge, ter

5 SYNTHESIZED SPEECH FOR EVALUATION OF CHILDREN'S HEARING AND ACOUSTIC-PHONETIC PERCEPTION

Mária Gósy

Phonetics Laboratory, Research Institute for Linguistics,
Hungarian Academy of Sciences

Key words: Synthesized Monosyllables, Auditory Screening, Acoustic Invariance,
Evaluation of Speech Perception

Abstract: *This paper describes the auditory screening function of low redundancy synthesized Hungarian words. They have successfully been applied to hearing screening of children. The same artificial words highlight the children's speech perception disorders in the case of normal hearing. The basic hypothesis is applicable to many other languages as well.*

1.   INTRODUCTION

The examination of hearing can be either objective or subjective. Despite methodological developments of objective auditory examinations (e.g., Sohmer and Feinmesser, 1967; Stockard and Rossiter, 1977; Hecox and Burkard, 1982; Eggermont, Ponton, Coupland, and Winkelaar 1991; etc.) the obtained results contain limitations that lead to the continuous need for subjective testing of hearing. There is a lack of an appropriate and reliable hearing screening procedure for use with young children. What are the criteria that a suitable method for auditory screening of small children (between the ages of three and seven) have to meet, taking into consideration the child's psychological and cognitive abilities? First, the sound signal that is given to a child's ear should be natural and familiar. Second, the measuring task should be easy to understand, that is, we should make it easy for the child to understand what he or she has to do during the testing. Third, the measuring method should yield the highest possible amount of information about the hearing mechanism operative between 200 and 8000 Hz. Fourth, the testing should be as quick as possible.

Hungarian speech synthesis results in recent years have made it possible to develop a new procedure that facilitates the solution of these problems (Olaszy, 1985; Olaszy, Gordos and Németh, 1992).

2.   THE BACKGROUND THEORY

The initial assumption was that sound sequences consisting exclusively of invariant features (or containing just a little more information than that) are the appropriate material for hearing examinations. The two prerequisites for the idea to work were that such sequences should be possible produce and that the way they sound should be equivalent to natural speech. A method based on such materials would, in a sense, encapsulate the merits of both pure tone audiometry and speech audiometry (Hazan and Fourcin, 1983). Natural speech is inappropriate for hearing examinations, except for speech audiometry testing, because of its high acoustic redundancy. However, speech is familiar to children, and repeating words would be an easy task for them. Synthesized speech contains less acoustic information than human speech does and is capable of providing information concerning pre-determined frequency ranges of the domain of hearing.

How is it theoretically conceivable that hearing capacity could be tested by means of synthesized, artificial sound sequences or words? The acoustic structure of natural speech is highly redundant, i.e., the speech signal contains significantly more information than is necessary for its accurate recognition (see, for example, Perkell and Klatt, 1986). Let us suppose that the mechanism of speech perception has to analyze the word *chair* that contains data of quantity $x+y$ where x represents the necessary (and sufficient) amount of acoustic data for safe recognition and y respresents the surplus, i.e., the amount of redundant elements. In the case of the example *chair*, the data surplus (y) becomes stored and can be called out immediately in case of any kind of disorder to provide supplementary information for speech decoding processing. For example, in the case of uncertain hearing or slight hearing loss, the stored information (y) will help the decoding mechanism to arrive at correct recognition. (The same applies to recognition under noisy circumstances.) What happens, however, if the speech signal, in our case the word *chair*, contains only the necessary amount of acoustic information (x) without almost any surplus (y)? The recognition of a sound sequence containing information x requires the processing of all information in a perfectly sound fashion, first of all, by means of normal hearing. Consequently, in the case of uncertain or impaired hearing, acoustic information of the quantity of x will not be sufficient for recognition, so the recognition process will be impaired and its result incorrect. Let us have a closer look at this point! If a person with a hearing impairment has to recognize a word on the basis of the acoustic information of x-z, where z is the lost acoustic-phonetic information as a consequence of impaired hearing, he or she will not be able to fulfill the task without the earlier stored acoustic-phonetic information of y. If there is no opportunity to use the stored redundant information, the recognition of the word will definitely be incorrect.

The first spectrogram in Figure 1 indicates the acoustic structure of the synthesized syllable [se:], while the others indicate its acoustic structures after various high-pass filtrations. High-pass filtration simulates hearing losses. The recognition of the syllable is based on the acoustic structures shown by the spectrograms. The threshold curves of the audiograms representing hearing losses are appropriate to the actual filtrations. Syllables at the bottom of the figure are the responses of persons with hearing losses (see audiograms).

On the basis of language-specific experimental results aimed at defining the speech sounds' primary and secondary acoustic cues for Hungarian (Gósy, 1989; 1992), specially synthesized speech could be developed. The artificially produced monosyllables could contain those acoustic parameters for each sound that had been set at pre-determined values without any acoustic redundancy. The decoding process could be simplified by activating only the speech perception of the decoding mechanism. It is known from perceptual studies that the consonant [s] in a sequence like [se:] is perceived correctly if the fricative has a noise component somewhere above 4000 Hz in terms of frequency. Hence, if this sequence is artificially produced such that the noise focus of [s] is at 5000 Hz but the frequency domain does not contain any other components that would suggest that consonant in the given period of time, the recognition of that sequence will be absolutely correct. Suppose, however, that the sequence is heard by someone with a hearing loss from 4000 Hz upwards. This person will have trouble at the first level of signal processing: depending on the degree of the individual's hearing deficiency, this person will identify the partly heard elements as different consonants like [ʃ, f, h] or some unvoiced stop. The identification of [se:] will be distorted into [ʃe:, fe:, he:, pe:] or [te:], and finally to [e:]. On the other hand, a naturally pronounced [se:], whether spoken by a male or female voice, will be perfectly identified – up to a certain degree of hearing loss – on the basis of additional (redundant) acoustic information of natural speech. The identification of the vowel [e:] – which also contains the most necessary acoustic elements, i.e., the first two formants (400 Hz and 2000 Hz for Hungarian) – depends again on the degree of the person's hearing loss at the frequencies in question. The recogniton of this synthesized palatal vowel by a person with a hearing impairment will result in various other vowels like [ø, o, u]. Misidentifications both of the consonant and the vowel demonstrate the place and extent of the person's hearing loss. Figure 1 shows the results of the first series of laboratory experiments for listeners with normal hearing and listeners with hearing impairments, using the example of synthesized sound sequence [se:].

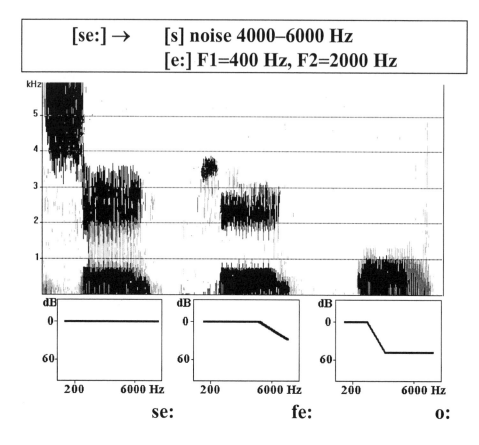

Figure 1.   Demonstration of the basic principle of the GOH method.

3.   THE PRODUCTION OF THE SYNTHESIZED WORD
MATERIAL

To provide a basis for the GOH method (the three letters of the abbreviation mean
the initial letters of the developers' names (Gósy and Olaszy) while H stands for the
word *hearing*), special test material was constructed which consisted of 60
meaningful monosyllabic Hungarian words. The criteria for choosing the words
were as follows: 1) the monosyllables should have two or three speech sounds
without consonant clusters; 2) the words should contain vowels and consonants for
which the frequency parameter serves as the primary acoustic cue for recognition;
3) the test material should include three types of items: words containing only high-
frequency sounds (like   [sy:z] 'virgin'), words containing only low-frequency
sounds (like [bu:] 'sorrow'), and words containing both high- and low-frequency
sounds (like [me:z] 'honey'); 4) most of the words should be familiar to children
of ages three to seven, however, the samples should also include a few words that
are meaningless for the children (i.e., unfamiliar); and 5) part of the test material

should contain words whose meaning could be expressed by pictures (in order to test easily also the children having speech defects). Regarding the criteria of 1) and 2), the following speech sounds were chosen as building elements for the words: [aː, a, o, u, e, eː, i, y] and [m, b, d, g, ʃ, s, ʒ, z, ts, tʃ, ɲ].

The process of formant synthesis was gradually built in several steps:

1. Tape recording the items of the test material in natural male pronunciation. (The speaker's F0-value was 120 Hz on average.)

2. Analysis of the acoustic structure of the recorded items.

3. Definition of the primary frequency cues for each speech sound.

4. Generation of the artificial equivalents by means of the Hungarian formant synthesizing system (called Univoice) using the primary acoustic cues.

5. Comparison of the perception of natural and synthesized words by means of listening tests.

6. Improvement of the sound quality of synthesized words by means of minor alterations of their acoustic structure.

7. Recognition of the synthesized words with the participation of both adults and children.

Table 1 shows examples concerning the used acoustic elements of some Hungarian speech sounds. The words were tape-recorded eight seconds apart; this duration of pauses between words was just enough for children to recognize the sequence and repeat it aloud. Attention was paid to the order of the words in the test material: low-frequency and high-frequency words alternated with one another. So, all children could have an experience of success, because they could recognize and repeat correctly at least every second word.

Table 1. Acoustic elements of some synthesized speech sounds.

Sounds (IPA)	Parameters		
	t (ms)	F1/N1 (Hz)	F2/N2 (Hz)
uː	120	250-300	600
iː	180	210-250	2200
eː	200	400	2000
ʃ	120	/1800	/2500
s1	120	/4000	/4000
s2	120	/6000	/6000
s3	120	/8000	/8000

(F stands for formants, N stands for noise, and s1, s2 and s3 represent three different [s] consonants.)

4. THE APPLICATION OF SYNTHESIZED WORDS FOR HEARING SCREENING

In the first series of experiments both the natural and synthesized word materials were filtered under *laboratory conditions* simulating various types of hearing losses with various extents. Adults and children with normal hearing listened to the words and had to identify them. Our working hypothesis had been that hearing deficiencies could be detected by means of specially synthesized words at a single constant intensity level. The results confirmed the hypothesized hearing screening function of the specially synthesized words since there was a significant difference between the percentage of correct recognitions of natural and synthesized words (p<0.0001). Table 2 shows examples of the data obtained. Percentage orrect recognition scores are given for five natural and synthesized words that were manipulated by means of low-pass and high-pass filtration with the cut-off frequency of 2000 Hz.

Table 2.   Percentage correct recognition of filtered natural and synthesized words.

Words	*Correct*	*recognition*	*of filtered*	*words (%)*
	Low-pass	*filtration*	*High-pass*	*Filtration*
	Natural	*synthesized*	*Natural*	*Synthesized*
meggy 'cherry'	100	14.16	96.66	70.00
síp 'whistle'	100	0	98.33	85.00
bú 'sorrow'	100	90.83	95.00	0
ász 'ace'	75.83	0	96.66	0
bot 'stick'	100	95.00	94.16	0

In the second series of experiments, the synthesized word material was used under *clinical circumstances* with the participation of 1000 children with hearing impairments (ages ranged from three to seven) and 500 children with normal hearing of the same ages (at the Department of Oto-Rhino-Laryngology at the Heim Pál Children's Hospital). Half of the children were girls while the other half were boys in both groups, and the IQ-level of all tested children fell into the normal range. The children with hearing impairments had hearing losses of various types and extent. Three kinds of examinations were carried out with all children: 1) pure tone audiometry; 2) tympanometry, i.e., impedance audiometry; and 3) hearing screening procedure by the GOH method with the synthesized words.

The synthesized words were administered to the children through headphones, one ear at a time. The intensity level was 45 dB on average in the case of children with normal hearing, while it was altered according to the threshold curves in the case of children with hearing impairments (the same procedure as is used in clinical speech audiometry). The intensity level of the synthesized words was controlled by clinical audiometer (type Beomat 905 DV). The children's task was to repeat each item immediately upon hearing it.

The 100% correct recognition of the artificial words at the intensity level of 45 dB on average means, in terms of pure tone audiometry, that the threshold curve runs between 0 and 10 dB, i.e., the person has normal hearing. If the threshold curve shifts downwards, i.e., there is some hearing loss, there will be less than 100% correct recognition at this intensity level.

5. RESULTS

Figure 2 shows that the incorrect responses for the synthesized words increase together with the increase of the intensity value of the threshold curve of the hearing-impaired child. Statistical measurements confirm a strong correlation between the correct recognition of the synthesized words and the intensity level of the threshold curve (r=0.76717).

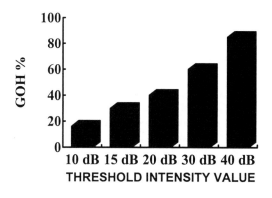

Figure 2. Interrelationship of incorrect responses to the synthesized words (%) and the pure tone audiometry values.

These results suggested the further idea that threshold curves might be estimated from word recognition scores. Early in the history of audiometry, Fletcher tried to draw conclusions concerning a person's speech comprehension from the data of the pure tone audiometry (1929). In this study the threshold curve was estimated from word recognition performance. This was made possible by the fact that the synthesized speech sounds making up the GOH word material contained frequency-oriented recognition cues between 250 Hz and 8000 Hz. The estimated threshold curves, made on the basis of the responses to 10 synthesized words in each ear, showed success in approximating the actual threshold curves obtained by pure tone audiometry with a 5 dB margin of error. As an example, a six-year-old boy's performance with his left ear is considered. For a detailed explanation, see Table 3.

The boy's correct performance is 40% of all responses, i.e., he could recognize correctly only four words out of ten (*meggy* 'cherry', *bú* 'sorrow', *bot* 'stick', and *bőr* 'skin'). General conclusions: the boy repeats lower-frequency words more exactly than higher-frequency words. The correct recognition of [b, m, o, u] suggests that his hearing is normal at 250, 500 and 1000 Hz. The incorrect recognition of [i:] and the unvoiced dental fricatives shows that the child has a hearing loss from 2000 Hz upwards. The fact that he failed to identify [i:] but correctly recognized the alveolar fricative consonants makes us conclude that his hearing starts to diverge from normal somewhere between 1000 and 2000 Hz. This claim is based on the following reasoning: in order to recognize [i:] the 2200 Hz second formant should have been recognized – which was not the case; the recognition of [ʃ], however, can rely on a 1500 Hz component. The recognition of

[s] shows that at the frequencies concerned (4000 Hz, or 6000 Hz or 8000 Hz for various dental fricatives) there is some loss in the child's hearing. The 4000 Hz noise in [su:] was recognized as [f], whereas the 6000 Hz noise in [se:l] was recognized as a plosive, and the 8000 Hz fricative in [a:s] was not recognized at all. These responses show a gradual increase in the child's hearing loss from 4000 Hz upwards, i.e., the latter recognition indicates that there was nothing of the original noise for the child to use for identifying the consonant. In sum, the course of the threshold curve is roughly as follows: up to 1000 Hz it appears to be normal, between 1000 Hz and 2000 Hz it decreases some 15 dB, and between 2000 Hz and 8000 Hz a further 20 dB. The pure tone audiogram of this child confirmed the estimated threshold curve. The calculated and the measured values of all tested children do not diverge in any significant manner. Consequently, the synthesized word procedure is applicable for the hearing screening of children of the tested ages.

Table 3.   A six-year-old boy's responses to synthesized words (bold letters are for words containing high frequency sounds, while italics are for words containing low frequency sounds).

Original	Synthesized	Words	Responses	of the boy
Orthographic form	IPA-form	Meaning	IPA-form	Meaning
Meggy	MeJ:	Cherry	meJ:	Cherry
síp	ʃi:p	Whistle	ʃut	Nonsense
bú	Bu:	Sorrow	bu:	Sorrow
ász	a:s	Ace	a:	Vowel
bot	Bot	Stick	bot	Stick
szél	se:l	Wind	te:l	Winter
méz	me:z	Honey	me:g	Yet
zsír	3i:r	Fat	3u:r	Party
bőr	bø:r	Skin	bø:r	Skin
szú	Su:	Woodworm	fu:	Nonsense

Comparisons of the children's subjective responses to the synthesized words were carried out with the objectively measured values of tympanometry (compliance and pressure). Tympanometry is a measurement of acoustic impedance as a function of air pressure variations, and reflects change in the physical properties of the middle ear system and tympanic membrane as air pressure in the external ear canal is varied. Middle ear disease is generally detected by an abnormal tympanometric pattern. For example, liquid in the middle ear, which is very frequent in childhood, reduces the normal mobility of the tympanic membrane and ossicular chain, and this results in a flat tympanometric configuration (Block and Wiley 1994). Figure 3 shows the interrelationship of the percentage of correct responses and the values of the compliance, where moderate correlations were found both in the cases of 50% or more incorrect responses ($r=0.52427$) and in the cases of less than 50% incorrect responses ($r=0.43666$). (The average value of the compliance in normal hearing with the used device is 1 or 1.2 ml.)

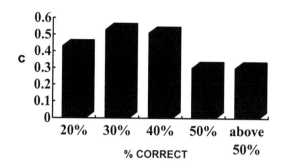

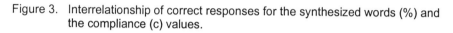

Figure 3. Interrelationship of correct responses for the synthesized words (%) and the compliance (c) values.

Figure 4 shows the pressure values of tympanometry for each performance level on the synthesized word task (% incorrect). Results confirmed again medium correlation (r=0.50174) between the children's responses and the objective values. Not a single child was found – out of 1000 – to repeat the synthesized words correctly but have a threshold curve not falling within the domain of normal hearing, or not have normal values of tympanometric measurements.

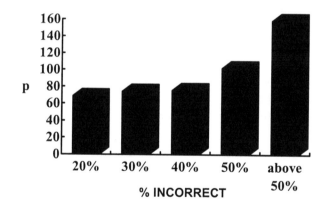

Figure 4. Interrelationship of incorrect responses for synthesized words (%) and the values of pressure (p).

5. 1. Results of the Screening Procedure

Further developments were planned for application of the procedure in a kindergarten setting. In order to solve all technical difficulties, a compact device was installed in a briefcase specially designed for the application of the synthesized material for hearing screening purposes in kindergarten. It included: 1) a switch for left ear/right ear selection; 2) a two-stage switch for two preset intensity values (an average of 45 dB for clinical and 55 dB for kindergarten circumstances, taking into consideration the basic noise level of the room in a kindergarten as usual in the hearing screening procedures); 3) a built-in cassette player providing constant transmission – like an audiometer – at any frequency (the transmission characteristic of the pick-up is corrected); and 4) a set of small headphones, the transmission characteristic of which meets hearing screening requirements.

On the basis of the clinical results for children having normal hearing and those with hearing impairments, an answer sheet was developed for hearing evaluation. The children's possible responses were categorized into four columns as shown in Table 4. The degree of hearing loss increases by approximately ten dB between columns. Responses of the first column appear in the case of normal hearing, and responses of the second column appear in the case of slight hearing loss, i.e., when the threshold curve decreases approximately ten dB. Responses of the third and fourth columns appear in the case of hearing losses when the threshold curves decrease a further ten or twenty dB. Patients having hearing losses more severe than 40 dB will not be able to hear anything from the synthesized words at the given intensity level. The Table 4 responses are written in IPA form with underlining used to indicate meaningful words in Hungarian.

Table 4.   GOH answer sheet for hearing screening (in IPA form).

Good hearing	Slight hearing loss	Hearing loss	Severe hearing loss
moʃ 'wash'	mos bos noʃ	Pus buf tuf <u>bot</u>	Pu, tu, up, ut, o, u, -
ʃiːr 'cry'	ʃiːl, siːl, <u>seːl</u>	Foːr, fuːl, <u>hiːr</u>	pu, tu, o, u, -
bab 'bean'	dad, bad, dod	<u>Pap</u>, pot, pup	tu, ut, u, o, -
syːz 'virgin'	syːs, siːz, seːs	ʃoː, ʃuː, fo, fu	to, tu, ot, ut, -
aːJ 'bed'	aJ, <u>ag</u>	aːd, oːg, oːd	tuː, <u>uːt</u>, o, u, -
søː 'weave'	syː, søːm, <u>syl</u>	<u>soː</u>, ʃoː, foː, <u>hoː</u>	to, tu, o, u, -
bor 'wine'	bol, bod, dor	Poː, puː, pup	to, tot, tu, o, u, -
tʃoːk 'kiss'	ʃoːk, ʃoːt, ʃoː	Foːk, <u>fut, hoː</u>	tot, to, tu, o, u, -
kuːt 'dwell'	tuːt, kuk, tut	Kuː, tuː, ut	u, -
eːs 'mind'	teːs, peːs, <u>meːs</u>	Eːʃ, eːf, øf	oːt, <u>uːt</u>, o, u, -

2400 children (600 three-year-olds, 600 four-year-olds, 600 five-year-olds, and 600 six-year-olds, boys and girls) were tested (both in the capital and the countryside of Hungary) by the GOH method under kindergarten circumstances using the answer sheets. None of them had any previous audiological testing. Each child heard ten words presented to the right ear and ten words presented to the left ear. Three different sets of words (altogether 60) were used in the experiments. The answer sheets were very simple to use. The examiner indicated the child's response by marking the appropriate item on the sheet and indicating the tested ear. Testing was begun with the right ear with half of the children and with the left ear with the other half of them, taking into consideration the right-ear advantage with the older children. No significant difference was found among the ten-word sets of the screening material, and no significant difference was found in the results depending on whether testing started with the right or left ear, either. Table 5 summarizes the results obtained by the GOH method in kindergartens according to the four categories. Children with evaluated hearing losses were sent to clinical examinations that confirmed their impairments.

Table 5.   Hearing screening results in kindergartens.

Evaluation of hearing	Ratio	of	children	(%)
	3	4	5	6
	(age	in	years	)
good hearing	77.83	82.16	89.50	93.33
slight hearing loss	13.5	14.5	9.66	5.83
hearing loss	5.83	2.16	0.66	0.66
severe hearing loss	2.83	1.16	0.66	0.16

5.2 Evaluation of Acoustic-Phonetic Perception

Clinical and kindergarten test data also allowed us to evaluate the children's acoustic-phonetic perception in the case of normal hearing. Those children who made an effort to comprehend words familiar to them in all cases (instead of repeating the words without understanding them) and/or misidentified the quantity of the speech sounds in the words were hypothesized to have difficulties with normal acoustic-phonetic operations of their speech perception process.

Phonetic analyses of all the responses of 500 normally hearing children showed that distinctions between slight hearing disorders and perception disorders clearly could be drawn. Four categories have been defined in order to summarize the children's perceptual mistakes: 1) misidentification of the quantity of speech sounds like [meJ:] → [meJ] or [sy:z] → [syz]; 2) deletion of a sound of the word that has no hearing screening function like [se:l] → [se:] or [ʃi:r] → [ʃi:]; 3) insertion of a speech sound into a word like [e:s] → [ke:s], [moʃ] → [moʃt]; and 4) voiced/unvoiced uncertainty of fricatives like [ø:s] → [ø:z] or [3i:r] → [ʃi:r]. Since perceptual disorders lead to difficulties of reading and spelling acquisition (cf. Gósy, 1997), it is worth testing the children's age-specific perceptual operations as well. Significant perceptual disorders were found in 41.11% of all 3-year-old children, 30.02% of all 4-year-old children, 22.53% of all 5-year-old children, and 13.92% of all 6-year-old children. (Considering these results, the older children were sent to further speech perception and comprehension testing.)

6.   CONCLUSIONS

Synthesized speech has many planned uses throughout the world. The project of medical application within the domain of hearing screening has been successfully established and used in Hungary. Low redundancy artificial words provide an opportunity to control the human hearing process, and this can be used in auditory screening of young children, where results should be obtained quickly but reliably.

Laboratory, clinical and kindergarten results have supported our hypothesis concerning the hearing screening function of specially synthesized artificial words. In addition, the same synthesized words highlight the children's speech perception disorders (concerning their misidentification of the given monosyllables) in the case of normal hearing. Speech perception disorders definitely are hypothesized to occur in children with hearing impairments because the impaired hearing hinders the appropriate operation of the perception mechanism. Although the actual data for the synthesized words and the results of the hearing screening procedure discussed in this paper are language-specific, the basic hypothesis is applicable to many other languages as well.

REFERENCES

Block, M. G., and Wiley, T. L. (1994). Overview and basic principles of acoustic immittance measurements. In J. Katz (Ed.), *Handbook of clinical audiology* (pp. 271-283). Baltimore: Williams and Wilkins.

Eggermont, J.J., Ponton, C.W., Coupland, S.G., and Winkelaar, R. (1991). Frequency dependent maturation of the cochlea and brainstem evoked potentials. *Acta Otolaryngologica, 111*, 220-224.

Fletcher, H. (1929). *Speech and hearing in communication.* New York.

Gósy, M. (1989). *Beszédészlelés.* Budapest: MTA.

Gósy, M. (1992). *Speech perception.* Frankfurt: Hector.

Gósy, M. (1997). Speech perception and comprehension of dysphasic children. In W. Ziegler, and K. Deher (Eds.), *Clinical phonetics and linguistics* (pp. 29-37). London: Whurr Publishers.

Hazan, V., and Fourcin, A. J. (1983). Interactive synthetic speech tests in the assessment of the perceptive abilities of hearing impaired children. *Speech, Hearing, and Language, 3*, 41-57.

Hecox, K., and Burkard, R. (1982). Developmental dependencies of the human brainstem auditory evoked response. *Annals of the New York Academy of Sciences, 388*, 538-556.

Olaszy, G. (1985). *A magyar beszéd leggyakoribb hangsorépítő elemeinek szerkezete és szintézise.* Budapest: Akadémiai Kiadó.

Olaszy, G., Gordos, G., and Németh, G. (1992). The Multivox multilingual text-to-speech converter. In G. Bailly, C. Benoit, and T. R. Sawallis (Eds.), *Talking machines: Theories, models and designs* (pp. 385-411). Amsterdam: Elsevier.

Perkell, J.S., and Klatt, D. H. (Eds.). (1986). *Invariance and variability in speech processes.* Hillsdale, NJ: Lawrence Erlbaum.

Sohmer, H., and Feinmesser, M. (1967). Cochlear action potentials recorded from the external ear in man. *Ann. Oto-Rhino-Laryngologica, 76*, 427-435.

Stockhard, J., and Rossiter, V. (1977). Clinical and pathologic correlates of brainstem auditory response abnormalities. *Neurology, 27*, 316-325.

6 UNIVERSAL ACCESS AND ASSISTIVE TECHNOLOGY
John C. Thomas, Sara Basson, and Daryle Gardner-Bonneau*

IBM T.J. Watson Research Center
*Michigan State University/Kalamazoo Center for Medical Studies

Key words: Universal Access, Assistive Technology, Human-Computer Interaction, User Interface Design, Augmentative and Alternative Communication (AAC)

Abstract: *Speech technologies have been a blessing to many people with disabilities. They have allowed people with severe physical impairments to do meaningful work, blind people to access computer technology, and people with speech impairments to communicate, for example. This chapter champions the concept of universal access – employing technologies in designs that serve both those with disabilities and those without. It also discusses the ways in which speech technologies are currently being used in assistive devices, and the problems associated with the current technology. Finally, the authors describe how methodologies and techniques from the disciplines of human-computer interaction (a.k.a. user interface design, usability engineering, and human factors engineering) can be used to better design applications to serve people with disabilities and the population at large.*

1. UNIVERSAL ACCESS VS. ASSISTIVE TECHNOLOGY

The concept of universal design implies designing products that are usable by everyone, including people with disabilities. "Curb cuts" are often cited as a classic example of universal design principles. Wheel chair users lobbied extensively in order to for legislation for curb cuts at all street corners, which allowed wheel chairs to roll easily from one side of the street to the other. Years after curb cuts were the norm, a sample of passersby was polled informally about the purpose of curb cuts. Respondents offered a number of reasons that curb cuts were introduced, such as access for shopping carts, bicycles, and skates. Wheel chair access was not the most frequently cited reason. A design feature – in this case, curb cuts – enhanced usability for *many* groups, *in addition* to users with disabilities.

Universal design remains an ideal, not easily achieved. Even the classic curb cut example is not a universally usable solution: blind pedestrians who depend on identifying the edge of the curb to safely cross the street have been hampered by curb cuts. Also, enhanced features that make products usable by everyone may increase the price of these products, thereby making the product less usable for people who cannot afford it. Additional "flexibility" will increase product complexity, typically – again, limiting usability for some individuals.

Vanderheiden (1997) neatly presents the distinctions between "universal design" and "assistive technology." Assistive technology refers more specifically to products designed for a particular user or disability group. Assistive devices tend to be more expensive than "universally designed" products, since the additional cost is distributed across a smaller population. This will also limit general usability, since not every potential user will be able to afford the custom-designed product. Also, as technology advances, assistive devices most in need of technological enhancements may lag behind, since they do not benefit from economies of scale as do mass market products. Finally, by making these products unique, individuals who need assistive devices may nonetheless shun them, to avoid being stigmatized as different. Indeed, one of the best arguments for universal design, though we don't like to admit it, is that people without disabilities fear people with disabilities, and everything associated with them (e.g., assistive technologies) that is unique or different (Covington, 1998). In fact, a poll conducted in 1991 by Harris & Associates showed that 47% of those interviewed admitted to feeling fear in the presence of an individual with a disability, and 58% indicated that they felt anxious, uncomfortable, or embarrassed in that situation. Universal design can help to increase the extent to which people without with disabilities can lead normal lives, with the same independence and privacy enjoyed by those without disabilities.

A strict universal design focus presents advantages as well as disadvantages, as does a strict focus on assistive technology. Vanderheiden advocates a combined approach. There will still be a need for assistive devices, even as universal design principles become more pervasive. On the other hand, standard product designers can often build in accessibility in early product phases, at dramatically lower costs than there would be if these products were retrofitted after the fact. Indeed, for some interfaces, accessibility can only be possible if it is built with the product; it may be "impossible to glue accessibility onto the systems as an afterthought or post-manufacture process." [Vanderheiden, 1997, p. 10]

Speech technologies present a dramatically different mode of interaction than non-speech interfaces. Compare the typical menu-driven touch-tone interface to an interface driven by speech. A touch-tone menu must explicitly enumerate all options: "For service, press one; for repairs, press two; for information, press three." A speech-driven interface can mimic this and state, "For service, say one; for repairs, say two; for information, say three." Most designers would agree that this presents an inelegant speech interface that fails to capitalize on the ease of use which speech can provide. Since speech responses need not be mapped to their touch-tone equivalents, users can instead be asked: "Do you want service, repairs, or information?" Users can respond by stating their choice. The fundamental difference between touch-tone and speech argues for incorporating both technologies early in the application design process. Customer responses, and even the course of the dialogue, will be different with each. Using speech as an "afterthought" add-on will result in a less attractive application, for all users, whether or not they have disabilities. Building speech into the early design phase will ensure a more elegant design, and more nearly meet the goal of universal access.

2. PREDICTED ENHANCEMENTS AND IMPROVEMENTS TO UNDERLYING TECHNOLOGY

Progress continues in the performance of speech recognition and speech synthesis systems. To some degree, this progress is an outgrowth of the steady improvements the computer industry continues to generate in processing speed and storage for a given cost point. Progress in these areas, in turn, makes it more feasible to base speech recognition systems on larger amounts of training data and on more contextualized telephone models. Similarly, synthesized speech can be constructed from a larger and larger repertoire of better contextualized models.

In addition, there is some room for optimism that additional performance improvements in speech technology can come from better understanding. For instance, current recognition systems do not do a very good job of speaker adaptation or adaptation to changing environmental background noise. It would seem that there is no theoretical reason why computer systems could not be developed that do a much better job of adaptation and hence better "front-end" telephone recognition. Current ASR systems also rely, typically, on statistical models of word sequences rather than explicit syntactic, semantic, and pragmatic analysis. Yet, the hope remains that at some not-too-distant time in the future, such considerations can be added to statistical models to give better "back-end" performance. Improvements in these areas might also provide more natural-sounding text-to-speech.

Such improvements mean that systems that are currently not feasible will become feasible for new points in the complex space defined by personal abilities, context, task, and cost. However, in some cases, improvements in non-speech technologies will also have such an enabling effect. We will outline, next, some of these predicted advances in non-speech technologies that will interact with speech technologies and then discuss possible new applications.

2.1   Pervasive Computing

Today, we tend to think of a computer as something that sits on a desktop. Yet, there are already more microprocessors embedded in other kinds of devices than are embedded in a distinct object called a "computer." As storage and processing costs continue to drop, we may begin to imagine an intelligent matrix of intelligent processors. These can be embedded in "wearable" devices such as non-obtrusive headphones, displays inside glasses, sensors and effectors inside gloves, and wristwatch computers and radios. We can expect computational devices to appear eventually in paper, cars, doors, windows, wallets, purses, pens, shelves, and plugs. These devices will communicate with each other (e.g., with infrared or low power spread spectrum radio) and thus be connected into the World Wide Web. We might sometimes carry active "badges" that signal who we are and where we are to this pervasive computational matrix.

2.2   Intelligent Materials

Eventually, even the material out of which our clothes, tools, and furniture are made will be able to sense environmental stimuli and respond.   There is already a snowboard whose flexibility varies in response to environmental changes. Eyeglasses whose darkness varies with light intensity have been available for some time.   We can imagine clothing for which the insulation properties change depending on sensed cold, casts for which strength and porousness change in response to pressure, and walls that can vary their sound insulation properties as our needs for socialization and concentration change.

2.3   Knowledge Management

A critical part of being able to handle intelligently the plethora of information that will become available as computation devices become embedded into the fabric of the world will be tools that help us deal effectively with knowledge throughout the entire "life-cycle" of creating, communicating, disseminating, finding, and learning knowledge. We expect to see systems that allow us to better handle semi-structured information such as stories and pattern languages; systems that help us identify, formulate, and solve problems; systems that help organize huge amounts of information and present it in ways that are easy to comprehend for particular purposes; and systems that help us get in touch with people who have the knowledge to help us in specific situations.

2.4   Architectural and Informational Integration

There is increasing emphasis at the research labs at MIT, Gesellschaft für Mathematik und Datenverarbeitung mbH (GMD), and Xerox on integrating architecture with information technology (see, for example, Streitz, 1998; Streitz and Russell, 1998; Underkoffler and Ishii, 1998). For example, imagine that chairs have embedded displays within them. They broadcast their location and the display tilt to the room. When two or more chairs are pulled together, the displays are linked; the lighting in the room changes to enhance the group awareness and to prevent screen glare.

The four technological trends mentioned above are not themselves speech technologies. However, together with speech technologies, we believe that they make certain applications more useful and usable. For example, as you walk around, the intelligent matrix knows who you are and where you are. If speech is one of your preferred modes of putting information into a computer, then your speech templates follow you around and are downloaded into any device with which you need to interact. Alternatively, you could carry your templates with you and "badge in" to any device that you need to use. For instance, assume that you are a user without sight. You might carry a badge that would, on voice command, alert you with auditory signaling to the location of the nearest terminal to the Internet.

Once you were ready to use the terminal, you could "badge in." This would alert the terminal to use voice input/output, the terminal would load your voice templates so that speaker-dependent recognition could be used, and the billing for Internet access would be carried out according to your wishes as expressed on the card. Furthermore, the system could then use voice verification to ensure that the terminal was really being accessed by you.

If you carried a headphone system, it might be in communication with the plan of the building. This could alert you auditorily to any potentially dangerous situations in the building. Of course, knowledge management systems would have to be in place so that any changes to the building plans would be registered.

Such systems would also be quite useful to sighted individuals. For instance, in a typical spontaneous meeting today, even a small number of participants makes automatically recording and transcribing of the oral contents of a meeting virtually impossible. Even listening to a tape recording of the meeting (without stereo sound) becomes onerous. However, if each participant had his/her own recording device that communicated with the others, a central source could combine the audio tracks to make a multi-threaded conversation. We could imagine transcribing such individual tracks (using speaker-dependent technology) much more easily. In addition, automatic summaries could be generated from the text. Some time later, you could access this summary. When you came upon a particular subject of interest, you could access the associated (admittedly imperfect) transcription. However, you might well be able to retrieve the information you wanted. If not, you could use the transcription as a guide to access the original speech recording and listen to that.

Hearing aid technology continues to advance, but hearing aid wearers still struggle in "cocktail-party" settings. Their ability to separate a target conversation from background noise is impaired. An intelligent system might sense the direction that a person with a hearing impairment is facing, and selectively amplify the sounds emanating from that direction, effectively filtering out extraneous ambient noise.

Generally, we tend to imagine ways that new technology can be used to allow differently-abled people to join in wider communities. One of the examples above, for instance, outlined how a combination of pervasive computing and speech technology could allow easier access for a person without sight to the Internet. However, we can also imagine using technology to support a virtual community with special needs. To some extent, this is already happening on the Internet. There are many existing listservers and web pages for people with various kinds of special needs.

3.   CURRENT ASSISTIVE TECHNOLOGY APPLICATIONS EMPLOYING SPEECH TECHNOLOGY

Although the previous paragraphs give us a peek into the future concerning the direction in which technology is moving and the types of applications that may result, they do not reflect the current uses of the technology by people with disabilities. In some cases, the degree of integration of applications and technology is nowhere near the level portrayed in the previous section. In this section, the current use of speech-enabled assistive technologies is described, along with the current problems users face with existing applications.

3.1   Applications Employing Automatic Speech Recognition (ASR)

One of the early assistive technology applications of speech recognition was in Environmental Control Units (ECUs).   Originally envisioned as a glitzy convenience for up-scale homes (i. e., "smart" houses), they were embraced quickly by the assistive technology community. "Butler in a Box," (Mastervoice, Los Alamitos, CA) was one very popular exemplar of this application. For people with severe physical disabilities, ECUs afford significant increases in independence, because they allow users to control their appliances, their heat, their lights, and other technologies used in the home, decreasing the amount of time that personal assistants are needed.   But ECUs have not been well designed in every case. Parsons, Terner, and Kearsley (1994) conducted a study with the purpose of improving remote control units of environmental control devices. They found that a large percentage of older subjects made errors in operating the units.   When manufacturers of the devices were queried about their consideration of the elderly in the design of these products, none of the ones that responded indicated that they took the needs of this population into consideration.   Given that the needs of the elderly, a considerably large population of users, was not considered, it is doubtful that other groups with special needs have been considered, either.

One of the speech technology applications that has received the most positive attention from the assistive technology community and the media is the use of ASR for computer access and control. For people with repetitive strain injuries (RSIs), who can no longer type for long periods on a keyboard, and for people with severe physical impairments who can not use their hands, ASR systems for dictation and command and control sometimes mean the difference between having a job and not having one. The ability to do computer work at home makes employment possible for some individuals who have never had the opportunity to be employed before.

Nevertheless, people with disabilities using speech recognition face the same usability problems with the technology that everyone else does, and then some. Engineers define quality of assistive devices in terms of a number of factors, including function, durability, reliability, robustness, manufacturability, ease of repair, and interactive control of "smart" assistive technology (i.e., interoperability, application integration) (Cooper, 1998).   ASR systems remain problematical with respect to at least two of these factors – robustness and interactive control. They are still very "fussy." ASR systems often don't work if a user has a cold, and sometimes don't adjust well to environmental noise conditions. They are picky when it comes to the sound cards and microphones with which they are compatible. Furthermore, despite marketing hype, speech recognition software is not well integrated with many of the software applications it is intended to control. All one needs to do to verify this is to monitor some of the many speech recognition discussion groups on the Internet. Though advertisements may say you can "run any software application by voice," those advertisements just aren't true at this point in time.   In many instances, users have trouble controlling even the specific applications called out by the product vendors in their own advertisements. Users are often in the position of having to develop their own "fixes" or "workarounds" to achieve their goals. Their diligence in doing so is a testimony to the enormous potential of this technology to better the lives of people with disabilities.

Speech recognition, in conjunction with text-to-speech (TTS), also has potential utility for people whose voices are unintelligible to listeners (e.g., some individuals with cerebral palsy). Edwards and Blore (1992), for example, have shown that as long as the speaker is consistent in his or her pronunciation of words, a speaker-independent ASR system can often recognize the speech more accurately than a human listener.

Finally, two other notable problems exist with current speech recognition technology. First, a sizable proportion of potential users of ASR systems requires *totally*-hands-free control, and not all of the systems provide this capability. Second, although ASR is being used, increasingly, as a replacement for keyboarding by individuals who suffer from RSIs, there are growing reports that extensive use can cause a condition analogous to RSI in the vocal cords.

3.2   Applications of Synthetic Speech

Screen readers have been an enabling technology for blind computer users for many years. However, with every new technology development, problems have been created for the functioning of screen readers. Screen readers functioned quite well for text-based, command-line interfaces (MS DOS and Unix). These interfaces were followed, however, by graphical user interfaces (GUIs) that created all sorts of problems related to screen navigation (See Mynatt and Edwards (1995) for a description of some of these problems and non-visual alternatives to GUIs.) Just when some of these problems became solvable, the Internet and World Wide Web became popular, creating even more problems for blind users, as web sites employed all manner of images, animation, flashing lights, and other visual devices that wreaked havoc with screen reader technology. Interestingly, one Internet application – e-mail – did become accessible to the blind, but primarily because there was a push in the general population for the capability of having e-mail read via synthetic speech over the phone. Harry Blanchard and Steven Lewis, in Chapter 12, describe some of the challenges faced in developing this service and some of the design decisions that were made. E-mail delivery represents a good example of a case in which a design intended to provide a convenience for the population at large provided ready access for the blind computer user to what was previously a very unusable application.

The good news with respect to the World Wide Web is that most web sites easily can be made 80-90% accessible through appropriate text tagging of their graphical elements. The bad news is that, despite the fact that there are programs to check web sites for accessibility (for example, BOBBY, from CAST, Inc., Peabody, MA) and books written about accessible web design (e.g., Waters, 1997), many web developers remain ignorant of the needs of blind computer users and/or unwilling to make the small investment required to make their sites usable. Alan Newell (1995) suggests that ignorance of the needs of the assistive technology field is the likely explanation for the current problem.

Whether the problem is ignorance or willful neglect, however, we are now seeing just how much impact the Americans with Disabilities Act (ADA) (1990) and other legislation are having in forcing change. Designers of public web sites can expect to be sued if their sites are not accessible, just as the city of San Francisco was recently sued for having inaccessible public kiosks. The plaintiff won that case, and the city is barred from installing any more kiosks until the ones already installed are made accessible. It is also required to take several other actions, including education of city employees about accessibility issues.

The field of augmentative and alternative communication (AAC) has also made use of synthetic speech in electronic devices for non-speaking individuals and individuals whose speech is unintelligible to listeners. Users of this technology can type input into, essentially, dedicated computers, and that input is then converted via TTS technology into speech for the listener. In the case of young children and other non-readers, special symbol systems such as Bliss or multi-meaning icons (e.g., Minspeak (Prentke Romich Company, Wooster, OH)) are used, the symbol input being translated into words and sentences and spoken by the synthesizer. For those with additional physical disabilities that do not allow them the fine motor control required for typing, these devices can be equipped with single- or multiple-switch scanning capabilities.

AAC technology is very complex and, not surprisingly, can not always be used effectively by the individuals for whom it is intended. First, evaluating clients to determine whether this technology is even suitable for them is difficult. The speech and language specialist must be able to assess the client's expressive and receptive language skills, symbol-recognition skills, and physical ability to access the device (Galvin and Caves, 1995). Second, even if there is a fit between the client's skills and the capabilities of a device, there is no guarantee that the client will be able to use the device in the various environments where it is needed. There are many barriers to use of these devices.

Equipment abandonment is one of the most serious problems faced in the assistive technology field. On average, one-third of all devices are abandoned, mostly within the first three months of use, and abandonment rates have been estimated to range from 8% to 75% (Scherer and Galvin, 1997). There are many reasons for abandonment of assistive technology, and most of them seem to apply in the case of AAC devices. Some of the issues reflect human factors problems:

1. Lack of training. Many of the electronic AAC devices are quite complex and require significant amounts of training. This training is not always available; nor is it always well designed.

2. Service and maintenance problems. Devices are abandoned because they are too difficult to maintain and service.

3. Design don't meet users needs.

4. Equipment viewed as stigmatizing or unacceptable by the intended user.

Intelligibility of the synthetic speech is not the only determinant of acceptability of AAC systems (Mirenda, Eicher, and Beukelman, 1989). If you have to rely on an AAC device to communicate, the synthetic speech from that device is, effectively, your voice. Consequently, it must be able to convey not only information, but emotions, and it must be viewed as consonant with the user's personality. Murray, Arnott, Alm, and Newell (1991), from the Microcomputer Centree at Dundee University in Scotland, successfully demonstrated a system called "Hamlet," that added recognizable emotions to synthetic speech via the manipulation of prosodic features. Nevertheless, most of the devices in use today do not readily convey the emotions of the user. The synthetic speech used in these devices, though better than it used to be, still sounds mechanical. That fact alone is enough for some users to reject the devices. Clearly, work remains to be done in this area to improve the prosody of synthetic speech, and the ease with which devices allow users to convey moods and emotions. As Cook and Hussey (1995) note, the incorporation of prosodic features can make these systems more acceptable to users, and can also make synthetic speech easier to listen to for long periods of time. Because high-quality synthetic speech systems that incorporate prosodic features are available today, one is led to believe that cost may be one factor limiting their use in current AAC devices.

4. HUMAN–COMPUTER INTERACTION: DESIGN AND EVALUATION

Effective use of speech technology requires an approach that takes into account the user, the task, the context, and the technology. Such an approach is necessary, in general, but also happens to lend itself quite well to the application of speech technology to universal design principles.

There are a wide variety of techniques for designing successful human-computer interaction. These include: checklists and guidelines, user-centered design, formal modeling techniques, heuristic evaluations, ethnographic study, field testing, and laboratory usability testing. While all of these techniques tend to be useful in various situations (Thomas, 1995), considerable difficulties exist in extending the results obtained with some of them to real contexts and real users trying to perform their real tasks (Thomas and Kellogg, 1989). In the case of dealing with new technologies and users with special needs, we believe that a combination of ethnographic study and field testing is most suitable, because these techniques are carried out in context.

Nielsen (1994) found that heuristic evaluation was most effective when the evaluators were familiar, not only with the tasks, and general human-computer interaction issues, but also with the specific technologies. We might well imagine that in the case of trying to perform a heuristic evaluation for people with special needs, it would also be useful for the evaluator to have experience with the specific needs. For similar reasons, checklists, models, and even laboratory techniques would have to be designed specifically for the particular complex of context, users, and tasks. If one does not use field testing, the combination of uncertainties concerning the projection of various other techniques into the real world multiply, and the chances of producing a system that real users would actually find useful becomes small.

Of course, the difficulty with field testing is that it implies one is fairly far along in the development of a product. Hence, many decisions may have been made already, and changing those decisions can be prohibitively expensive. There are several ways to mitigate this potential problem. One is that the system can be built on a flexible architecture, and tools can be provided for rapid prototyping and change. Another is to spend time really understanding the tasks, the users and their needs, and the context before even formulating the problem to be solved by the design. In addition, one can use incremental technological instantiation. By incremental technological instantiation, we mean that the form factors and the functions of the proposed technology may be gradually introduced without actually building the final technology. A familiar example of this is the "Wizard of Oz" trial in which a human being simulates a speech recognition engine with various kinds of capabilities and dialogue structures. Only after field testing reveals the appropriate technological specifications and user interface design does prototyping a real system begin.

To take another example, suppose you wish to test the idea of giving people who cannot hear a specially outfitted laptop computer. You can begin by describing the functions that you would hope such a computer would eventually have and give a potential user a blank box, essentially, with the same form factors to carry around in a real context and try to perform a real task. Then, you work together to attempt to see how and when such a device might be used and begin to develop the details. You can also discover some of the difficulties that such a device might pose in various real contexts. For example, it might be too heavy, it might get stolen, it might be the case that there is no convenient way to recharge the batteries, or it might be that workplace illumination requires a special screen, and so on. By having an actual object (even a non-functional one) incorporated into the person's daily life, you can discover potential design dilemmas prior to heavy intellectual (and financial) commitment toward a particular course of action.

Continually refining the ideas and carefully observing the results might be termed "organic research." The problems being addressed arise from real situations and the results are applied back to that real situation. But new knowledge, methods, or technology must be created in order to solve the problems. In creating new knowledge, methods, or technology, the researcher also draws on other domains to help visualize what is possible.

For instance, consider the example of relay chat for individuals who cannot hear. In the way this procedure currently works, a call can be initiated by either a hearing or non-hearing party. A relay chat operator (RCO) is called. The hearing person says something to the RCO, who then types what is said on a terminal. This is echoed on the terminal of the person who doesn't hear. Now, that person reads what is on the terminal, thinks about a response, and types it. This appears on the RCO's terminal, and he or she proceeds to read what is there to the hearing person.

This is a very lock-step form of communication compared with the way interactive speech usually operates. In fact, for interactive problem solving, Chapanis (1975) found that oral speech was an extremely efficient method of communicating, but only when the "interruptability" of speech was preserved.

When participants explicitly were forced to take turns, the superiority of speech over typed text disappeared.

In thinking about how relay chat might be improved, we can look at another situation that involves a similar set of constraints -- chat rooms and "instant message" facilities on the Internet, e.g., on America-on-Line (AOL). Here too, people are not speaking, and the communication is necessarily "turn taking." However, what people actually tend to do in most cases is have multi-threaded asynchronous conversations.

This conversational ploy allows "interruptability" in written communication at the cost of a greater load on the user's working memory. This must be a tradeoff that is preferred, judging from the high level of activity on AOL in the deaf community. The point here is that in designing an automated relay chat based on ASR and speech synthesis, one need not limit the space of design possibilities by trying to "replace" what the human RCO does today. Instead, by examining other technologies (in this case, Internet communications), an automated RCO could be designed to operate in a "duplex" mode.

Speech technology contributes to universal design by creating an access mechanism that appeals to both people with and without disabilities. The state of the art in speech technology and other unrelated technologies, however, limits the ability of speech interfaces to create truly universal access. The development of futuristic technologies such as pervasive computing, intelligent materials, and knowledge management, coupled with enhanced speech technology, opens up avenues of access to all potential users. Because these services are novel, however, their design must be carefully explored, considering the user, task, context, and technology through human-computer interaction and human factors engineering methods, tools, and techniques. For populations with disabilities, this organic research process is even more critical, since designers may not be able to depend on their own expertise as sample users.

REFERENCES

Americans with Disabilities Act (ADA) (1990). PL101-336, 42 USC-12111.

Chapanis, A. (1975, March). Interactive human communication. *Scientific American, 232,* 36-49.

Cook, A. M., and Hussey, S. M. (1995). *Assistive technologies: Principles and practice.* St. Louis: Mosby.

Cooper, R. A. (1998). Incorporating human needs into assistive technology design. In D. B. Gray, L. A., Quatrano, and M. L. Lieberman (Eds.), *Designing and using assistive technology: The human perspective* (pp. 151-170). Baltimore: Paul H. Brookes.

Covington, G. A. (1998). Cultural and environmental barriers to assistive technology: Why assistive devices don't always assist. In D. B. Gray, L. A. Quatrano, and M. L. Lieberman (Eds.), *Designing and using assistive technology: The human perspective* (pp. 77-88). Baltimore: Paul H. Brookes.

Edwards, A. D. N., and Blore, A. (1992). Speech input for persons with speech impairments. In W. Zagler (Ed.), *Computers for handicapped persons* (Proceedings of the 3rd International Conference on Computers for Handicapped Persons) (pp. 120-126). Vienna: R. Oldenbourg.

Galvin, J. C., and Caves, K. M. (1995). Computer assistive devices and environmental controls. In R. L. Braddom (Ed.), *Physical medicine and rehabilitation* (pp. 493-501). Philadelphia: W. B. Saunders.

Harris, L., & Associates. (1991). *Public attitudes towards people with disabilities.* Washington, DC: National Organization on Disability.

Mirenda, M., Eicher, D., and Beukelman, D. R. (1989). Synthetic and natural speech preferences of male and female listeners in four age groups. *Journal of Speech and Hearing Research, 32,* 175-183.

Murray, I. R., Arnott, J. L., Alm, N., and Newell, A. F. (1991). A communication system for the disabled with emotional synthetic speech produced by rule. *Proceedings of Eurospeech '91: 2nd European Conference on Speech Communication and Technology* (pp. 311-314). Geneva, Italy.

Mynatt, E. D., and Edwards, W. K. (1995). Metaphors for nonvisual computing. In A. D. N. Edwards (Ed.), *Extra-ordinary human-computer interaction: Interfaces for users with disabilities* (pp. 201-220). Cambridge, UK: Cambridge University Press.

Newell, A. F. (1995). Extra-ordinary human-computer interaction. In A. D. N. Edwards (Ed.), *Extra-ordinary human-computer interaction: Interfaces for users with disabilities* (pp. 3-18). Cambridge, UK: Cambridge University Press.

Nielsen, J. (1994). *Usability engineering.* Boston, MA: Academic Press.

Parsons, H. M., Terner, J., and Kearsley, G. (1994). Design of remote control units for seniors. *Experimental* Aging *Research, 20,* 211-218.

Scherer, M. J., and Galvin, J. C. (1997). Assistive technology. In S. Kumar (Ed.), *Perspectives in Rehabilitation Ergonomics* (pp. 273-301). London: Taylor & Francis.

Streitz, N. A. (1998). Integrated design of real architectural spaces and virtual information spaces. *CHI '98 conference summary* (pp. 263-264). New York: ACM.

Streitz, N. A., and Russell, D. M. (1998). Basics of integrated information and physical spaces: The state of the art. CHI '98 conference summary, (pp. 273-275). New York: ACM.

Thomas, J. C. (1995). Human factors in lifecycle development. In A. Sydral, R. Bennett, and S. Greenspan (Eds.), *Applied Speech Technology* (pp. 289-314). Boca Raton: CRC.

Thomas, J. C., and Kellogg, W. A. (1989, January). Minimizing ecological gaps in interface design. *IEEE Software,* 78-86.

Underkoffler, J., and Ishii, H. (1998). Illuminating light: An optical design tool with a luminous-tangible interface. *CHI '98 Conference Proceedings* (pp. 542-549). New York: ACM.

Vanderheiden, G. (1997). Universal design vs. assistive technology. *Proceedings of RESNA '97.* Arlington, VA: RESNA Press.

Waters, C. (1997). *Universal web design.* Indianapolis, IN: New Riders Publishing.

7 GUIDELINES FOR SPEECH-ENABLED IVR APPLICATION DESIGN

Daryle Gardner-Bonneau

Michigan State University/Kalamazoo Center for Medical Studies

Key words: Interactive Voice Response (IVR), Menus, Anthropomorphism, Point of View, Guidelines, Structured Dialogue

Abstract: Although guidelines and standards have been developed for the design of the user interface for interactive voice response (IVR) applications, most of these documents consider only menu-driven, touch-tone-based IVR application designs with little, if any, attention being given to IVR systems that employ speech recognition and/or natural language dialogues. This chapter presents the history of user interface guidelines development for IVR applications and discusses two issues - menu-driven user interfaces and anthropomorphism - to illustrate how guidelines for speech-enabled IVR user interfaces may differ from those for touch-tone-based IVR designs. In addition, current guidelines specific to speech recognition-based IVR applications, incorporated in a document currently being developed for the American National Standards Institute by the HFES-200 Committee of the Human Factors and Ergonomics Society, are described.

1. INTRODUCTION

This chapter is written from the viewpoint of someone who has served as a human factors consultant to IVR application developers for about ten years and worked on developing guidelines and standards for IVR application design for the past five years. During that time, I've had the opportunity to observe what was happening in the field, and to be involved in solving some thorny problems. My colleagues and I have struggled with the tradeoffs and dilemmas that arise while developing guidelines for applications. The vast majority of IVR applications in the United States still employ the touch-tone phone for input. Thus, it is ironic that my first consulting experience in 1989, thanks to the forward thinking of Anita Bounds (then at American Express Travel-Related Services), was for an application that would employ an automatic speech recognition front-end. American Express deserves credit for being ahead of its time; speech recognition technology had not yet reached the point where its deployment in IVR applications was feasible. But as a consequence of this first experience, I never became wedded to the idea that IVR applications had to be touch-tone-based and menu-driven. This experience broadened my approach to the many IVR applications I subsequently helped develop. Although speech recognition may not have been ready for mass application in 1989, breakthroughs during the past two years indicate it might now be ready. The technology has improved dramatically, and we are now seeing more and more IVR applications making use of it.

In the late 1980s, IVR applications became ubiquitous, and a wide variety of applications - some good, some not - were foisted on the public. Application developers who needed to design successful applications (or improve failed ones) were clamoring for human factors assistance. They lamented the fact that there were no guidelines or standards to help them. Why did this situation exist? First, IVR application design has never been a serious topic of academic research in the way that speech recognition and synthetic speech are. Thus, research data and papers providing guidance for IVR design were not coming out of the academic community. Second, telecommunication companies that were doing the lion's share of human factors research in this area were reticent to share data with potential competitors. They published papers, some quite useful, but shared limited data when it came to specific design data (e.g., the number of seconds appropriate for an input time-out following a prompt to the caller). Their work was proprietary. One notable exception to this was the speech technology group at Ameritech. It was exceedingly generous in sharing its knowledge, even going so far as to post its phone-based user interface standards and design guidelines (Schwartz and Hardzinski, 1993) on its web site. Third, while telephone companies have a long history of human factors research and development activities, many of the people clamoring for guidance were not associated with telephone companies. IVR applications were being developed and deployed by many businesses and organizations that had never heard of the term "human factors engineering." Therefore, developers relied on whatever assistance they could get from equipment vendors and whatever knowledge they had acquired based on experience with other IVR applications.

There was one group of people that recognized the seriousness of the situation posed by the lack of standards and guidelines. These individuals were involved with voice messaging/voice mail applications, where the lack of standards constituted a real burden to the user because so many different user interfaces were being implemented. Many users had to interact with a variety of voice messaging systems. They used their own systems at work, at home, and in their cars. They used even more systems when they called someone and wanted to leave a message on that person's system. Because users were faced with a myriad of inconsistencies from one application to another, it was more difficult to use voice messaging than it should have been. In the latter part of the 1980s, the Voice Messaging User Interface Forum (VMUIF) was formed to develop standards and guidelines for such applications. They produced a document (VMUIF, 1990) and later contributed to a standard (ISO/IEC 13714 - User interface to telephone-based services: voice messaging applications, 1995) that laid out consensus guidelines for the user interface design of these applications. A little more than one year later, an ANSI version of the standard followed (ANSI/ISO, 1996). Harry Blanchard, in Chapter 12, presents much of the information contained in these documents.

Unfortunately, many, if not all, existing voice messaging systems violated these guidelines. Could they, or would they, be brought into compliance? Some developers probably never heard about the guidelines, and thus never attempted to comply with them. Developers and suppliers that knew of the standards were faced with the problem of client resistance to changes to a system that had already been learned. (How many people have switched to Dvorak keyboards after becoming proficient with QWERTY keyboards?) John Chin (1998) recently presented a paper outlining his experiences in attempting to apply the standard in retrofitting an application developed before the standard was in force. His conclusion was that some changes were possible, but others simply weren't.

Although the standards were suitable for voice messaging IVR applications, it was not clear they could be applied effectively to other IVR applications. Specifically, voice messaging is a common application with a substantial base of "power users." It does not represent the many custom IVR applications that callers encounter when they telephone, for example, a government agency, a travel service, or their local electric company. Furthermore, the standards were limited to touch-tone-based interfaces that were menu-driven. Recognizing both the limitations of these standards and the fact that developers might never encounter these documents, I began to push for the development of a general IVR design guidelines document in the early 1990s (Gardner-Bonneau, 1992a). The intent was to develop a set of guidelines that could be distributed widely to application developers and their clients in business, industry, and public service (e.g., government agencies, foundations). Unfortunately, finding a group of people to develop such a document was difficult. The people who really needed the document didn't have the human factors expertise to produce it. In addition, the effort required a sponsor - a home base from which development of the document could occur. It wasn't until 1996 that the effort found a home - with the Human Factors and Ergonomics Society (HFES). There existed within HFES a standards committee known as HFES-200 that, for a number of years, had been providing input to Parts 10 through 17 of ISO 9241 (Ergonomic requirements for work with visual display terminals (VDTs)). Parts 10 through 17 of ISO 9241 provide guidelines for the design of office system software user interfaces. (Parts 1 through 9 provide guidelines for hardware user interfaces, primarily Visual Display Terminals (VDTs).) This committee was being encouraged by the HFES to develop an ANSI version of ISO 9241, Parts 10-17, as well. The committee agreed to do so, and submitted a Project Initiation Notification (PIN) to ANSI. The PIN stated that the ANSI document would incorporate ISO 9241, Parts 10-17, and extend that document in three areas: 1) voice input/output and telephony; 2) accessibility of computers to people with disabilities; and 3) use of color in software user interfaces. The PIN was accepted, and the HFES-200 committee has been working on this standard since 1996, with a significant amount of committee time being devoted to the three sections of the standard that extend ISO 9241, Parts 10-17.

The Voice Input/Output and Telephony section of the standard was divided, initially, into four subsections: 1) speech recognition (both command and control and dictation); 2) speech output; 3) non-speech audio; and 4) interactive voice response. Recent major technology leaps in speech recognition have created significant difficulties in writing the command and control portions of the speech recognition subsection of the document. Technology developments in speaker-independent and large vocabulary speaker-dependent recognition (including dictation systems) have already rendered some of the command and control guidelines obsolete or unnecessary. The portion of the speech recognition subsection related to dictation, however, is fairly well developed, as it was drafted more recently. The speech output subsection has had very little attention to date, but will be relatively short. Early on, the committee decided there was insufficient information available to provide useful guidance in the area of non-speech audio. Task-relevant use of non-speech audio (including, but not limited to, earcons) in both computer and telephony applications is still in its infancy. Non-speech audio has not been exploited nearly to the extent that it could be, although some individuals in the field (e.g., Balentine, 1997) have demonstrated how non-speech audio could be used in task-relevant ways. Nevertheless, it was decided to drop the non-speech audio section, at least for the first release of the draft. The remaining subsection on IVR system design is the most complete, and may be released through the ANSI canvassing process within the next 12 months.

When we started work on the IVR subsection of the document, applications were still largely touch-tone based with menu-driven user interfaces. Not surprisingly, then, the current draft of the IVR subsection reflects this reality. However, attention also has been given to speech recognition-driven user interfaces. Much of the rest of this chapter is devoted to the issues we encountered in the development of this draft and a discussion of the implications of recent developments in speech recognition for IVR design guidance. Before turning to these issues, however, it is important to note one other recently initiated effort to develop user interface design guidance for speech recognition-driven IVR systems.

In the summer of 1998, a group of people calling itself the Telephone Speech Standards Committee was formed in the hopes of preventing a repetition of the situation described previously with respect to voice messaging. The group consists of developers of speech-enabled IVR applications and their associated vendors, and includes representatives from companies like Nuance Communications, Fidelity Investments, MBNA America, Syntellect, and Comverse Network Systems. One of its goals is to develop user interface guidance for speech recognition-driven IVR applications used primarily for business transactions. It was my pleasure to give a presentation (Gardner-Bonneau, 1998) to the group at its first face-to-face meeting. This meeting was sponsored by the American Voice Input/Output Society (AVIOS) at its annual conference. In that presentation, I outlined the work of the HFES-200 committee, as well as some of the issues and pitfalls of developing guidance in this domain. It is not yet clear how the work of this group will evolve over time, or what guidance they will be able to produce, but it is an effort worth watching. More information about the Telephone Speech Standards Committee is posted at the AVIOS web site (http://www.avios.com).

The remainder of this chapter outlines some of the persistent issues in IVR user interface design that created problems in developing design guidance. Particular attention is given to menu design and to the issue of anthropomorphism. Following these discussions, guidelines specific to speech recognition-driven IVR systems will be described in brief.

2.   MENU-DRIVEN IVR APPLICATIONS

Menu-driven IVR applications are, almost exclusively, an American phenomenon, and a pervasive one. Because touch-tone penetration is nearly 100% in the United States, U.S. applications have been driven by the constraints and limitations, as well as the strengths, of touch-tone technology.

When the input modality allows only the entry of ten digits and two other characters (# and *), the menu-based dialogue is the only option, at least the only option that is not unwieldy. Europe, in contrast, has far fewer applications of this type, given its much lower level of touch-tone penetration. It's almost as if Europe has skipped an evolutionary stage. It has focused on other aspects of speech technology, such as automatic translation and the processing of natural language. I would submit that menu-driven IVR applications will never be prevalent in Europe, and that speech-based IVR applications will become prevalent in the future. Furthermore, I predict that many, if not most, of those applications will *not* be menu-driven.

One of the first areas of significant disagreement with respect to drafting the standard was the use of the term "menu." In my thinking, a menu is, generally speaking, a poor metaphor with respect to IVR applications, simply because menus, in the traditional usage, are both visible and static. That very fact can lead developers, unwittingly, to treat an auditory "menu" like a visual menu, with the end result that they ignore many of the limitations of the auditory modality that must be considered to achieve good IVR application design. For example, since the auditory "menu" is not visible, the user cannot refer back to it easily. This frequently adds a memory burden to an auditory task that is not present in a visual task. In those instances, limiting the number of items in a menu is necessary. In addition, the issue of navigating auditory "menus" to refer back to previously presented menu items must be considered. A visible menu can allow users to see how an application is structured, greatly facilitating the navigation process. Such is not the case with "auditory" menus. Difficulty in navigation often leads users to what has been called "voice mail jail" (Greve, 1996), where they sit, hopelessly lost in an application, until they blindly find their way out, or hang up. I eventually won the battle to remove the word "menu" from the document. However, I probably lost the war, in that we still had the problem that most of our guidelines assumed a menu-driven interface, whether we were calling it a menu or not.

I took the extreme position on this issue, not because it was correct, necessarily, but in an attempt to emphasize the fact that, with the recent developments in speech recognition technology, we were no longer constrained to menu-based interfaces. Our guidelines needed to consider other types of dialogue designs. But are there occasions in which the menu-driven interface is still the best option, even with a speech recognition-based front end on the application? The answer is probably "yes," but the conditions under which this is true have yet to be elucidated. The chapter by Hansen and Novick in this volume (which outlines a task-oriented approach for determining when IVR applications are appropriate) may hold some promise for answering this question. There do appear, however, to be some obvious applications in which menu-driven designs may be optimal. Highly overlearned applications (e.g., voice messaging), may be most efficient, ultimately, as a consequence of having a menu-based design. In such applications, users are likely to become completely familiar with the menus as a result of extensive use, and can bypass menus through devices such as dial-ahead and dial-through. But is a menu-based design most efficient for other applications that are not subject to extensive use, are less fixed in their procedures, and are encountered casually by callers? Probably not.

2.1   Depth, Breadth, and Length of Menus

Determining the depth, breadth, and length of menus has been the subject of many papers and conference presentations (e.g., Virzi and Huitema, 1997). However, specific quantitative guidelines have not been set, and for good reason. Decisions about depth, breadth, and length of menus are heavily application-dependent and almost always involve a number of design tradeoffs related to the logical task structure of the application, limitations of human memory, and ease of application navigation. In the early 1990s, many applications failed, or were irritating to callers, because of: 1) navigation problems created by menus that were too deep; 2) menus that were too broad (creating the perception of wasting the listener's time); and 3) memory burdens due to overly long menus. Those who offered explicit guidance (e.g., Gardner-Bonneau, 1992; Schumacher, Hardzinski, and Schwartz, 1995) to help developers avoid the most common errors were criticized, to some extent, for overgeneralizing. Both Gardner-Bonneau and Schumacher et al. recommended that menus consist of four or fewer items, as a general rule. For many applications, this may be good advice. On the other hand, the case is made easily for exceptions to this rule.

For example, let's assume that the theatre owner of a local Cinema 8 wants to set up an IVR application to provide people with the show times for the feature films showing on the eight screens. The dialogue might operate like this:

System:	Welcome to Cinema 8. For titles and show times of the movies currently playing on our screens, press 1. For other information, press 2.
Caller:	*1*
System:	To hear show times for "Jurassic Park," press 1.
	For "The Wizard of Oz," press 2.
	For "The African Queen," press 3.
	For "Armageddon," press 4.
	For "The Rugrats Movie," press 5.
	For "Silence of the Labs," press 6.
	For "It's a Wonderful Life," press 7.
	For "Primary Colors," press 8.
	To hear these listings again, press 9.
	To return to the main menu, press "*."

If one follows the design advice rigidly, this menu should be split in two. But is that better? It's doubtful. The reasons why this menu can work, despite its length, are relatively simple, but important. First, it's likely that the person calling either already knows what movie he or she wants to see, or wants to find out both what movies are playing and when. If the choice of movie has been made already, the caller can discard from memory all menu options except the one in which he or she is interested. Also, because the items are short, the application cycles rather quickly through the menu. If the caller has not yet picked a movie, he or she can cycle rapidly through the menu and acquire information about one or more of the movies that are playing. The splitting of this type of menu into two would probably complicate what should be a relatively simple application.

3. APPLICATIONS WITHOUT MENUS

Does the movie theatre dialogue presented previously represent the optimum design by which the caller can achieve his or her goals? Perhaps. But some might argue that it would be simpler if a natural language dialogue employing speaker-independent speech recognition was available. The caller could just say the name of the movie for which show times are desired. Such a dialogue might work like this:

System:	Welcome to Cinema 8. For which movie would you like show time information?
Caller:	*The Rugrats Movie.*
System:	The Rugrats Movie is showing at 3:00, 4:45, 6:30 and 8:15 on screen 5.

This looks simple, and it is - deceptively so. But suppose the caller's response had been:

Caller:	*[to his children playing in the room] What's the name of that movie you want to see? [Then, into the phone] Oh, yeah, The Rugrats Movie.*

The system now must have some mechanism for filtering out the extraneous speech and for identifying the relevant words (i.e., "The Rugrats Movie") in the caller's speech. Techniques like "word spotting" can be used for this purpose, but probably not with perfect accuracy in every case. Furthermore, what if the caller doesn't want information about a particular movie, just the list of what's playing, or is calling for some other reason? The touch-tone-based dialogue handles all three situations fairly smoothly. Can a natural language dialogue using speech recognition be designed to achieve the same result just as efficiently? Does it need to? If 99% of the callers already know what they want to see, this simple dialogue may actually suffice, with a few adjustments to handle the exceptions. But if only 50% of the callers fit into this category, 30% want information about what's playing, in general, and 20% are calling for some other reason, a natural language dialogue to serve this situation could become quite complex.

One of the lessons to be learned from this is that it's critical to know the goal (or goals) of the callers. If incorrect assumptions are made about them, no one's needs will be served.   Colleen Crangle (1998) provides a variant of this "lesson." She likens the creation of an IVR dialogue to designing a work of fiction, in which such literary elements as *point of view* are extremely important.   Your point of view determines, to a considerable extent, what you'll see and not see, recognize and not recognize, be aware of and not, understand and not. Part of good IVR dialogue design is to understand the caller's point of view, and the impact that it should have on your application's design.

As an example, I once worked with an IVR dialogue for reporting power outage problems to an electric company.   Clearly, the dialogue, as shown to me initially, was written from the point of view of a utility repair person. One of the prompts to the caller asked whether the line from the house to the utility pole was damaged. The only responses available to the caller were "yes" and "no." If I'm a caller reporting a power outage, presumably, my power is out. I don't have lights, and it may be dark outside. How can I possibly be expected to answer this question as posed to me in the IVR dialogue? Obviously, the dialogue developer did not "stand in my shoes" when he or she put pen to paper.

Susan Boyce has nicely laid out some of challenges of natural language processing (see Chapter 2). It's clear that applications using speech recognition and natural language processing, unless they are extremely simple, lie somewhere in the future. It's a giant leap from speech recognition to language understanding, and an additional giant leap from language understanding to communication through conversational dialogues.   Unfortunately, the many potential consumers of these technologies do not discriminate between speech recognition and the other two. Thus, callers into IVR applications assume and expect applications to be smarter than is technologically feasible. Though we haven't yet reached the point of using true natural language dialogues in applications, we can fulfill user expectations in more traditional IVR dialogues with attention to the human factors engineering issues.   Applications may not be intelligent, as defined in artificial intelligence parlance, but they can appear to be "smart."

David Attwater and his colleagues at British Telecom Laboratories (1997) have demonstrated, for example, how dialogues employing speech recognition can be highly structured, yet appear very natural to callers. One of the ways this is done is by programming the application to make full use of the speech information it gains during the course of a call. The information acquired early in the call can be used to determine how the dialogue proceeds later. As a result, the system can avoid giving prompts or menu options that would seem foolish to the caller given his or her prior actions. It can also "infer" the caller's intentions and "predict" prompts that would be appropriate based on those intentions. In effect, callers' responses in a speech-enabled interface often provide more information than the application uses, and certainly more than the caller's responses do in a touch-tone-based dialogue. By making use of the cues the caller gives through speech responses, the IVR system can appear to be "situation-aware" or "smart," even though conventional dialogue design mechanisms are used to achieve this end.

As an illustration of Attwater's ideas, let's say that someone called a travel reservations center that employed an IVR system with a conventional dialogue design, though it allowed speech recognition as input. That dialogue, without the structure Attwater suggests, would play "twenty questions" with the caller to acquire all of the information it needed to book, for example, an airline reservation. It might ask, separately, for the date the customer wants to fly, the airline, the time (hour and a.m./p.m.), the destination, point of origination, etc. Eventually, the questions all would be answered, and the flight would be booked.

The dialogue might work something like this:

System:	To what destination would you like to fly?
Caller:	*I'd like the 3:30 Delta flight to Pittsburgh.*
System:	On what date would you like to fly?
Caller:	*[says today's date]*
System:	What approximate flight time would you like?
Caller:	*3:30*
System:	Is that A.M. or P.M?
Caller:	*P.M.*

The job will get done, but it will take a while. But if we look at the caller's first response, we can see that he or she has explicitly provided several of the pieces of information needed to process the reservation. In addition, and just as important from Attwater's perspective, the caller *implicitly* has given the system one piece of information based on the way the response was phrased: the caller wants to fly *today*. If the system knows the current time (i.e., the time the call is made), it can also infer, with near certainty, whether the requested flight is an a.m. or p.m. flight. Rather than to ignore the information supplied, Attwater recommends that a structured dialogue contain fall-through mechanisms that would use the additional unprompted information, both implicit and explicit, to move the dialogue along more quickly. Thus, the system dialogue could proceed, as follows:

System:	To what destination would you like to fly?
Caller:	*I'd like the 3:30 Delta flight to Pittsburgh.*
System:	From what airport do you wish to fly?
Caller:	*Tampa.*
System:	Delta Flight 427 leaves from Tampa for Pittsburgh at 3:30 this afternoon and has seats available. Do you want to proceed with the reservation?
Caller:	*Yes, please.*

Note that if the system has the caller's current location available to it, it may also be be possible to infer the origination city, rather than having to prompt for it.

4. HYBRID APPLICATIONS

Given the extent to which multiple input modalities are employed in computers today, it seems fair to ask whether hybrid applications - those that combine touch-tone, often menu-driven input with non-menu-driven (most likely) speech recognition input - are potentially useful for telephony applications, at least in the near-term. In fact, considering the constraints of today's technology, such hybrid applications may actually constitute the most efficient means for callers to complete some tasks. One's first reaction might be that such applications would be confusing to callers. However, for anyone that uses computers, switching between technologies to accomplish a goal is a fact of life. If we can switch effortlessly between the mouse and the keyboard, we should be able to switch from speech to touch-tone input with similar ease. In order to do this, we need to assess the costs and benefits of modality switching, and then structure dialogues such that the most efficient use can be made of both input modalities. It remains to be seen whether such hybrid applications will be developed, and how common they might become.

5. ANTHROPOMORPHISM

One of the long-running arguments in IVR system design is the extent to which anthropomorphism should be employed in IVR systems. It might seem interesting that the argument persists despite strong pronouncements against anthropomorphism from Quintanar, Crowell, and Pryor, in 1982, which were later rearticulated by Schumacher et al. (1995). But then the book, *The media equation* (Reeves and Nass, 1996), published a matter of months after the Schumacher et al. paper, served to "stir the pot" again with its discussion of computers as having both gender and personality. As a consequence, in early 1996, an intense electronic discussion of this issue occurred, involving over 50 people who were subscribers to four electronic lists devoted to user interface design. What follows are some of the conclusions drawn from that discussion.

First, it should be noted that people had very different ideas about what the term "anthropomorphism" really means. Three definitions emerged during discussions: 1) using animate terms for inanimate items; 2) using gender terms for genderless items; and 3) using terms usually reserved for humans to apply to non-humans. According to Merriam-Webster's Collegiate Dictionary (1993), anthropomorphism is "an interpretation of what is not human or personal in terms of human or personal characteristics." To anthromorphize, according to the same source, is "to attribute human form or personality to things not human." Though we may disagree on some of the specifics, it's clear that we're talking about giving a machine a personality that it then exhibits via its speech in an IVR application.

For many, it seemed, the question of anthropomorphism was asked and answered with Microsoft's BOB application, an abject failure in terms of giving a machine a personality. Curiously, the negative attitudes of developers towards anthropomorphism persist to this day, despite the success, particularly with casual users, of the Wildfire automated attendant application (Wildfire Communications, Boston). As late as the summer of 1998, when I was involved in a teleconference call with human factors specialists in which this issue arose once more, many professionals viewed anthropomorphism as inappropriate. On the other hand, there was a sizable and vocal minority who espoused the opposite position.

5.1   What's Right and Wrong With Anthropomorphism?

Many of those who took the negative position felt that anthropomorphism was particularly obnoxious in voice, less so in text, user interfaces, even though the message content is the same in each case. "We thank you for shopping ABC" is not typically perceived as offensive in written form, but may be in a voice message, because it's patently clear that the speaker is a machine and not "we." Those individuals committed to direct manipulation interfaces felt that anthropomorphism wasn't needed, didn't work, and would never work. Predictability, comprehensibility, and control were perceived as the key user interface attributes by some in this group. They cited numerous examples of failed anthropomorphic interfaces. Finally, anthropomorphism was viewed as fundamentally dishonest by some in this group, because the machine is being represented as something it is not - a human being.

On the other side of the argument were those who were positively disposed towards anthropomorphism or felt that it might be appropriate in at least some cases. Not surprisingly, individuals working in the field of education and computer/electronic games development fell into this group. They maintained that anthropomorphism fails, when it fails, because it is badly implemented or implemented for the wrong applications. For example, anthropomorphism may only get in the way in the context of very simple applications. This group also felt that proponents of direct manipulation were overstating their case. For graphical user interfaces (GUIs), where the primary problem is visualization, their points might have some validity. On the other hand, in auditory interfaces, where the problems often concern communication and/or decision-making, direct manipulations solutions would be less successful. This group also pointed out that in multi-user interface situations some form of social interaction is often necessary to prevent communication breakdowns. Some also felt that if anthropomorphism could achieve the goal of decreasing the need for user attention to system navigation, it would be useful. Less attention could be paid to the interaction, and more could be paid to the task.

The argument that has seemed to move the masses from the extreme negative position to a neutral one, at the very least, is that the extent to which anthropomorphism is appropriate depends on the "corporate image" an organization wishes to convey through the application. In many large businesses, the formal, impersonal, business-like language of an application, with a minimum of anthropomorphism, might well be appropriate and most acceptable to callers. For small organizations, particularly those that, for example, specialize in custom personal services or education and outreach (as opposed to business transactions), an anthropomorphic application might well be appropriate. In short, if my client is AT&T, formality may be best. But if my client is Ben and Jerry's Ice Cream, Mary Kay Cosmetics, or Planned Parenthood, a different level of personalization might be desired, both by the client and the callers.

5.2   Anthropomorphism as a Usability Issue

Beyond pointing to failed applications that employ anthropomorphism, detractors had little in the way of usability data. Nor did proponents. The one exception was that some detractors noted that the use of the personal pronoun, "I", by the system in telephony applications occasionally caused callers to assume that there was a human being, rather than a machine, on the other end of the telephone line. This, in turn, led to confusion and communication breakdowns during the calls. This problem, however, can be avoided rather easily if "I" is not used until the application has firmly established, through the early parts of the dialogue, that the system is an automated one.

Therefore, there is some question as to whether anthropomorphism is a usability issue, or only a user preference and acceptance issue. It may well be that anthropomorphism is largely an acceptance/preference issue in touch-tone menu-driven, IVR dialogues in which there are many cues available to remind callers of the limitations of the technology. On the other hand, anthropomorphism becomes a usability issue as we move to IVRs employing speech recognition, natural language processing, or both, where such cues will be less obvious. Further, if one envisions calls in which there are multiple participants, and in which one or more of these participants is a machine, it may well be the case that anthropomorphism becomes an *essential* element of the user interface. For example, anthropomorphism may need to serve as the means by which machine participants are uniquely identified by the human participants throughout the course of a dialogue. In teleconferences, humans identify each other by voice characteristics, personality characteristics, and communication styles. How will humans identify machines in such a situation, if not by similar characteristics?

Anthropomorphism is not simply a function of the words that are spoken. It also involves the way in which the words are spoken, and the characteristics of the voice used to convey them. As designers, we can manipulate all of these elements in an attempt to design in, or to design out, anthropomorphism. However, it is the user who will decide, ultimately, whether the resulting application is acceptable or not, and usable or not. If we deliberately set out to design an anthropomorphic application, it's important that the entire application reflect this approach and that the system not reveal itself to be a dumb machine, particularly when problem situations occur. The Wildfire automated attendant was noted, previously, to employ a modestly successful anthropomorphic interface. Nevertheless, there have been anecdotal reports that when communication does break down between Wildfire and the caller, mild frustration can become exasperation in short order. This can occur because the caller's frustration is not reflected in the responses of the automated attendant, which continues with its sweet, "no worries" prompts and tone of voice, revealing its inability to empathize with the caller.

5.3   Guidelines for Anthropomorphism

The electronic discussions about anthropomorphism suggest three guidelines, or guideline areas, regarding anthropomorphism, though we may not know enough to write all of the actual guidelines at this point. The first set of guidelines concerns the conditions under which anthropomorphism would be appropriate in telephony applications. The second guideline is that applications should be anthropomorphic or non-anthropomorphic, but not a mixture of the two. Finally, then, there should be guidelines concerning the "rules of behavior" for anthropomorphic and non-anthropomorphic user interfaces. For now, the HFES-200 committee members working on the IVR subsection have decided not to address the issue directly in the current draft.

We may not have definitive guidelines at this point, but Bruce Balentine, who participated in the electronic discussions, eloquently described a goal concerning anthropomorphism that many may find appropriate:

> "Truly successful speech applications won't be accepted until they do use the kinds of behavioral devices that personify and 'awaken' the machine as modeled by the user. But they will still not be machines trying to be people (anthropomorphism). They will be machines acting like very good and professional machines that possess certain human traits (mechanoanthropomorphism)."
> [Personal communication, 1996]

6. CURRENT GUIDELINES FOR SPEECH-ENABLED IVR APPLICATIONS

It is worth noting that most of the HFES-200 committee members working on the Voice Input/Output and Telephony section of the document were associated with telephone companies. It should come as no surprise then, that the document, as it stands, was heavily influenced by their particular experience, and by both ISO 13417 and, to a lesser extent, the Ameritech document discussed previously. Almost all of the guidelines are written in the context of menu-driven, touch-tone-based IVR user interfaces, or are general in nature (e.g., requiring the caller to enter any given piece of information only once, unless re-entry is required for reasons of privacy, security or verification).

In general, nearly all of the guidelines in the document can be applied to IVR applications with a speech recognition-based interface, provided those applications are menu-driven. The document currently contains only two guidelines exclusive and specific to user interfaces employing speech recognition. One guideline designates that the word "say" be used (as opposed to "speak" or "enter"), when speech recognition is used as the input mechanism. The second says that if enrollment by speech is required in an application, the enrollment should occur in an environment similar to the one in which the application will be used, if possible. It should be noted that this guideline was meant to apply to applications in which speaker-dependent speech recognition is used, as well as to applications in which the speech enrollment is required as part of a security procedure (i.e., speaker identification and/or voice verification).

Two other guidelines in the document specify touch-tone system-based functionality, along with an analogous, speech recognition-based alternative. The first of these states that callers should be able to interrupt any prompt, unless it is critical that they hear the information contained in it (e.g., in the case of certain error tones or messages). The user input that constituted the "interruption" should be processed. This is designated as "dial-through" capability for touch-tone-based applications, and "talk-through" for applications employing speech recognition.

The second guideline states that callers should be able to enter information that allows the system to process a series of actions without presenting any portion of the prompts to the caller. In other words, callers should be able to enter several uninterrupted commands in a row, with no intervening prompts. This is designated as "dial-ahead" in touch-tone driven applications and "talk-ahead" in speech-driven applications. The text notes, however, that talk-ahead capability may not be feasible in some applications for technical or system performance reasons. This caveat alone suggests that menu-driven IVR interfaces, which can be very efficient when dial-ahead capability is provided, may not meet users' needs in a speech-recognition driven user interfaces that attempt to use talk-ahead. An entirely different user interface may be required to achieve the goals of dial-ahead in a speech-driven application.

The previous three paragraphs outline the only guidelines specific to speech recognition-based interfaces in the IVR subsection of the HFES-200 document. There are, however, additional guidelines in the document's subsection on command and control user interfaces for applications employing speech recognition. Some of these guidelines may be applicable in the IVR subsection, as well. This possibility is currently being explored. The HFES-200 Committee is aware that the IVR subsection of its document, in its current version, does not do justice to speech-enabled user interfaces. On the other hand, it is unclear just how much additional guidance can be written at this point, based on empirical data and industry experience with speech-enabled IVR applications today. Nevertheless, we are interested in adding guidelines, where possible. In addition, the Telephone Speech Standards Committee, discussed earlier in this chapter, is also launching an effort to develop IVR guidelines for speech-enabled IVR applications used for business transactions. I welcome contact with all human factors professionals who are interested in these efforts, or have data and information to contribute.

7.   CONCLUSIONS

Recent developments in automatic speech recognition (ASR) have made it feasible to deploy ASR in telephony applications. However, effective use of ASR in IVR applications is more than just substituting spoken words for touch-tone input in a menu-driven application. Other types of dialogue designs may be more appropriate than menus when speech input is used. In addition, other factors (e.g., anthropomorphism) may be more important for speech-based interfaces than they are for touch-tone interfaces. Current user interface design guidelines contain little that is specific to IVR applications employing speech input. Those involved in applications development in this area may be wise to consider developing design guidelines for telephony applications employing ASR for two reasons. The process of developing guidelines can help to improve the quality of applications currently under development. In the future, it may help to avoid the user interface consistency problems that users of voice messaging systems have faced since the late 1980s.

ACKNOWLEDGEMENTS

I would like to thank Eileeen Schwab, of Ameritech, for her careful reviews of several versions of this chapter.

REFERENCES

American National Standards Institute. (1996). *User interface to telephone-based services: voice messaging applications* (ANSI/ISO 13714). New York: Author.

Attwater, D. J., Fisher, J. S., and Greenhow, H. R. (1997). Towards fluency - structured dialogues with natural speech input. *Proceedings of the 16th Annual International Voice Technologies Applications Conference (AVIOS '97)* (pp. 147-158). San Jose, CA: American Voice Input/Output Society.

Balentine, B. (1997). Speech technology in its own right - shedding the baggage of the past [Abstract]. *Proceedings of the Human Factors and Ergonomics Society 41st Annual Meeting* (p. 309). Santa Monica, CA: Human Factors and Ergonomics Society.

Balentine, B. (1996). Personal communication.

Chin, J. (1998). Alternative user interfaces: A case study of voice mesaging [Abstract]. *Proceedings of the Human Factors and Ergonomics Society 42nd Annual Meeting* (p.1611). Santa Monica, CA: Human Factors and Ergonomics Society.

Crangle, C. (1998). Fictional elements in the design of telephone-based conversational interfaces. *Proceedings of the 17th Annual International Voice Technologies Applications Conference (AVIOS '98)* (pp. 179-192). San Jose, CA: American Voice Input/Output Society.

Gardner-Bonneau, D. J. (1998, September). *Standards efforts of the HFES-200 Committee.* Presentation to the Telephone Speech Standards Committee at the 17th Annual International Voice Technologies Applications Conference (AVIOS '98), September 14, San Jose, CA.

Gardner-Bonneau, D. J. (1992a, October). Human factors problems in interactive voice response (IVR) applications: De we need a guideline/standard? *Proceedings of the Human Factors Society 36th Annual Meeting* (pp. 222-226). Santa Monica, CA: Human Factors Society.

Gardner-Bonneau, D. J. (1992b). Human factors in interactive voice response applications: "Common sense" is an uncommon commodity. *Journal of the American Voice I/O Society, 12,* 1-12.

Greve, F. (1996). Dante's 8th circle of hell: Voice mail. *St. Paul Pioneer Press*, April 8th, p. 1A.

International Standards Organization [ISO]. (1994). User interface to telephone-based services: Voice messaging applications {ISO/IEC DIS 13714). Geneva: Author.

International Standards Organization [ISO]. (date varies by Part). Ergonomic requirements for work with visual display terminals (VDTs) - Parts 10-17. (ISO 9241, Parts 10-17). Geneva: Author.

Merriam-Webster's collegiate dictionary (10th ed.) (1993). Springfield, MA: Merriam-Webster.

Quintanar, L., Crowell, C., and Pryor, J. (1982). Human-computer interaction: A Preliminary social psychological analysis. *Behavior Research Methods and Instrumentation, 14*, 210-220.

Reeves, B., and Nass, C. (1996). *The media equation*. New York: Cambridge University Press.

Schumacher, R. M., Jr., Hardzinski, M. L., and Schwartz, A. L. (1995). Increasing the usability of interactive voice response systems: Research and guidelines for phone-based interfaces. *Human Factors, 37*(2), 251-264.

Schwartz, A. L., and Hardzinski, M. L. (1993). *Ameritech phone-based user interface standards and guidelines* (Release 1.0). Hoffman Estates, IL: Ameritech.

Virzi, R. A., and Huitema, J. S. (1997). Telephone-based menus: Evidence that broader is better than deeper. *Proceedings of the Human Factors and Ergonomics Society 41st Annual Meeting* (pp. 315-319). Santa Monica, CA: Human Factors and Ergonomics Society.

Voice Messaging User Interface Forum [VMUIF]. (1990, April). Specification Document. Cedar Knolls, NJ: Probe Research.

8 LIMITING FACTORS OF AUTOMATED TELEPHONE DIALOGUES

David G. Novick[1, 2], Brian Hansen[3], Stephen Sutton[4], Catherine R. Marshall[5]

[1]EURISCO, [2]Oregon Graduate Institute, [3]Oregon Heath Sciences University, [4]Fluent Speech Technologies, Inc., [5]CollabTech, Inc.

Key words: Dialogue, Telephone, Modality, Task

Abstract: *If computers are to engage humans in spoken-language dialogues over the telephone, then what kinds of tasks should these systems perform? What limitations do human conversants bring to the interaction? What are the problems associated with the computer's side of the conversation? How does the use of the telephone modality limit what can reasonably be accomplished? Finally, how are we to understand the factors that distinguish those tasks that are appropriate for implementation as automated telephone dialogues (ATDs) from those for which current technology is not yet ready, or those which will never be appropriate for ATD treatment? In this chapter, we explore the limitations imposed by the user, the system, and the telephony modality and present several perspectives for understanding interactive tasks. From these we derive a feature space of factors critical to the success of any ATD. We conclude by applying the analysis to a broader range of tasks, employing the identified features.*

1. INTRODUCTION

The promise of automated spoken-language systems raises the question of what sort of applications could employ this emerging technology most successfully. Spoken-language systems seem clearly valuable in the ubiquitous modality of the telephone, in which (aside from touch-pad entry) the voice is the only medium of communication. If computers are to engage humans in spoken-language dialogues over the telephone, then what kinds of tasks should these systems perform? What tasks will humans want to do over the phone that computers can carry out? What limitations does the telephone modality impose on human-computer interaction? To answer these questions, we consider the effects of limitations of the conversing partners (human and machine) and modality on the range of applications to which automated telephone dialogue (ATD) technology might be usefully applied.

While the roles of the computer, the user, and the communication modality they use to interact may be reasonably clear, the characterization of a *task* in human-computer interaction is not well settled. The nature of a task depends on the observer's view. For instance, the meaning of the activities that constitute the execution of the task depends on understanding the goals driving the activities (Suchman, 1987). Consequently, we consider various perspectives on the notion of task. As a result of our analysis, we are able to identify a feature space that characterizes the appropriateness of tasks for ATD systems. The dimensions defining the feature space are the humans, the computers, the telephone modality as a medium, and the tasks that comprise an ATD in use. Our analysis produced 42 features in this space that predict the likelihood of success of a proposed ATD. We find that the most useful areas of the feature space are those that relate to two or more of the dimensions (such as user x task) rather than a single dimension (such as user only). This suggests that those who intend to develop ATD systems should avoid simplistic reliance on individual characteristics of users or systems in isolation.

In Section 2, we will discuss the kinds of constraints imposed by the nature of human-computer interaction: the limitations inherent in the computer's side of the interaction, the limitations imposed by the modality in which the interaction occurs, and the limitations (or requirements) that human users of spoken language technology bring to the interaction. In Section 3, we will examine differing perspectives on how to characterize tasks with the aim of deriving those factors that distinguish good candidates for spoken-language technology. Section 4 provides a detailed presentation of the feature space, which can be used to assess the suitability of an application for development as an ATD. In Section 5, we conclude with a broader look at possible applications.

2.   CONSTRAINTS ON HUMAN-COMPUTER INTERACTION VIA TELEPHONE

We begin our analysis by considering constraints that are essential to human-computer interaction. These are relevant because, for ATD technology to be useful for a specific task, it must be shown that the task is "doable" within the limitations set. There are three main sources of constraint relevant to human-computer interaction: those pertaining to the computer (or the ATD), those pertaining to the modality across which the interaction takes place, and those pertaining to the human conversant. A modality can be understood as comprising multiple channels, with each individual channel supporting a single interaction path such as voice, graphics, or gesture, among others. The human-computer interaction process, at its highest level, is depicted schematically in Figure 1. The user and computer communicate via a modality that affords access to multiple channels. We will consider the constraints imposed by each of these elements (human, computer, modality and task) in the following sections. Our discussion of these constraints begins with a summary of characteristics of the technology of automatic speech recognition and then uses this information as background for an extended treatment of the implications of these characteristics for spoken-language applications.

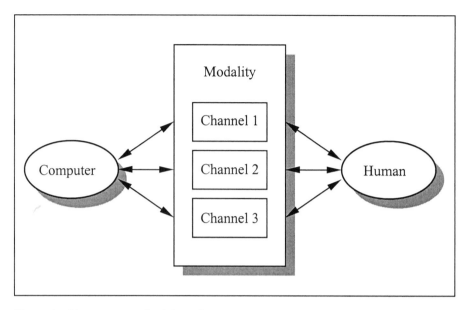

Figure 1. Human-computer interaction process.

2.1   Limitations of Computer Conversants

The most critical limitations of ATD technology are those associated with speech recognition capabilities. The biggest problems with speech recognition algorithms tend to be processing time, recognition accuracy, and lack of certainty for out-of-vocabulary words. Speech recognition researchers have made progress in each of these areas but problems still remain (Cole et al., 1995); ATD developers must make efforts to mitigate the effects of these problems. The problems are shared by both speaker-dependent and speaker-independent systems, although speaker-dependent systems tend to have better accuracy. This performance improvement, however, comes at the cost of extra training time and lack of adaptability to new users. As ATD systems typically involve a diverse population of users, they face problems of recognition accuracy beyond those of trainable, speaker-dependent single-user systems now becoming available for personal computers.

Recognition accuracy is also affected by the size of the recognition vocabulary. In general, increasing the number of words in the vocabulary reduces performance, in terms of both recognition accuracy and processing time. Since ATDs typically require real-time interaction, developers must ensure that the processing delay is kept within bounds deemed acceptable to the user. Recognition accuracy poses the greatest challenge: A system with a high potential for misrecognition would be confusing and frustrating to users. Moreover, it would be altogether unacceptable for a host of information-critical tasks.

A consideration that is closely related to the size of the vocabulary is the confusability of words in the vocabulary. In general, words that are similar in sound are more difficult to recognize than ones that are dissimilar. Furthermore, due to the nature of speech recognition algorithms, longer words tend to be easier to recognize than shorter words. This is both unfortunate and ironic because studies of the English language have shown that the relationship between word length and frequency is inversely proportional. That is, the most common words tend to be the shortest.

Finally, there is the issue of speech recognition robustness. Current systems often find it difficult to recognize when an out-of-vocabulary word is given as input. We have called this the rejection problem. One implication of this problem is that ATD designers wanting to avoid misrecognizing out-of-vocabulary words will tend to increase the size of the recognition vocabulary, straining the accuracy and speed of recognition.

Given these limitations of current technology, several considerations must be taken into account in order to produce effective ATD applications. The most important of these considerations is the necessity of producing and using predictions as to what a human user of an ATD is likely to do. These predictions can be applied at all levels of the conversational model, from phonetic to pragmatic.

It is not enough to simply make predictions, however; ATDs must make *good* predictions in order to function effectively. Consequently, it is often necessary to impose constraints on the range and freedom of users' responses. In this way the system can form more accurate expectations as to what users will say next. There are a number of ways of encouraging users to adapt their behavior (Brennan, 1991; Zoltan-Ford, 1991). *Entrainment* refers to a form of passive conditioning in which users adapt their behavior in line with the style of prior presentations by the system—this phenomenon is applicable at all levels of discourse including lexical, syntactic, semantic, and dialogue structure (however, see Kennedy, Wilkes, Elder, and Murray, 1988). Another method of constraining user input involves explicitly expressing to users the requirements for their input. For example, the system prompt "Would you like me to repeat anything? (please say yes or no)" gives explicit instructions to the user about how to respond (Hansen and Novick, 1996).

At a more general level, users' behavior can be manipulated via conversational *initiative*. Loosely speaking, initiative refers to the extent to which conversational participants can influence the dialogue in terms of the underlying purpose of what is next uttered (see Walker and Whittaker, 1990; Kitano and Van Ess-Dykema, 1991; Smith, 1994). While every utterance in a dialogue serves some purpose, the notion of initiative concerns broader interactive goals otherwise known as the discourse purpose (Clark and Schaefer, 1989). Dialogues may be single-initiative in which only one participant is responsible for establishing the discourse purpose, or may be mixed-initiative in which either of the participants may act to establish the discourse purpose. Typically, initiative is not something that is negotiated; rather it is a product of the form of utterances and of the specific task. An example of a user-initiated dialogue is one in which the user arrives at certain information by asking questions of the system. In contrast, in a system-initiated dialogue, the user may determine the same information by navigating through menus and specifying only one of a small set of choices. By choosing the appropriate types of utterance, it is possible for the system to assume the initiative from the beginning and to maintain it for the course of the dialogue. In this view, initiative is essentially a matter of degree—the degree to which the user is unconstrained. There are certain advantages to the system holding the initiative. Most important is the ability to form expectations as to what the user will say next. Although this may still be possible in a mixed-initiative case, the expectations are likely to be broader and less accurate.

Despite the intuitive appeal of initiative as a factor in discourse, the notion of initiative remains, in many ways, ambiguous or imprecise. For example, in human-human interaction, repairs may be initiated by either party within the context of an exchange initiated by one of the parties. Or one of the parties might initiate a side-sequence that is structurally similar to a repair sequence but is in fact not a repair of the utterance in focus. Regardless of definition, however, to the extent that conversants can undertake unanticipated repairs, side-sequences and tangents, a dialogue system may encounter difficulties in relying on initiative to constrain responses. Where initiative is maintained by the system, constraint becomes more practical; for communicative actions that continue a current exchange, the concept of initiative remains useful.

The main concern with attempting to influence and modify users' behavior is that, from the users' standpoint, such constraints tend to reduce the overall naturalness of interactions. In human-human conversation, people initiate repairs and side sequences as needed. Human conversants do not, for the most part, consciously rely on effects like entrainment; dialogue conventions are implicitly negotiated on the fly (Clark and Wilkes-Gibbs, 1986). Thus, a major challenge for researchers is to design dialogues that constrain the kinds of responses that the user may give, while maintaining a relatively natural and intuitive interaction. As ATD developers, we seek opportunities to explore and influence users' responses in light of users' natural expectations about interaction.

We note that this sort of approach, where the system relies on the natural ability and tendency of human conversants to adapt to their partner's conventions, places the entire burden of adaptation on only one party to the conversation. This asymmetry is, in itself, an "unnatural" condition because, pathological cases aside, both conversants mutually adapt in normal conversation. This suggests that there are limits to the naturalness of human-computer dialogue. Technological advances in recognition accuracy will enable computers to become adaptive by reducing the need to constrain user behavior so severely. In the meantime, "natural" dialogues are those constructed to maximize the illusion of freedom of action by the user— presumably through clever anticipation of users' responses.

In designing system outputs that best constrain user inputs, system developers need to address the perplexity of expected inputs. In avoiding prompts that produce highly variable responses, the problem is not necessarily open-ended queries (asking when the user was born as opposed to asking for each of birth month, birth date, and birth year, for example), in and of themselves. There could be open-ended queries for which we generally expect to get only one or two different answers or requests. Rather, whether the query is open-ended or closed, the chances of accurate recognition decrease as the number of possible responses increases. This is determined by the degree of constraint offered by the context of the interaction. Thus an open-ended question in a highly constrained context (due to task expectations, for example) might work well; but a closed-end question in a relatively unconstrained context is likely to be unworkable. Subtle differences in the way a question is phrased or delivered can affect the degree to which a human responds in phrases or with single-word answers, or the extent to which filled pauses ("uhm" or "uh") are used as pre-utterances.

In general, there is a tension between what might be called the "effectiveness" and the "naturalness" of ATD systems. The concept of effectiveness includes considerations such as recognition accuracy, processing speed, and robustness, while the concept of naturalness is associated with usability and vocabulary size. Recognition accuracy is inversely related to the confusability of the recognition vocabulary, while confusability generally grows with vocabulary size. Figure 2 depicts the relationship between effectiveness and naturalness, showing that advances in speech recognition technology will increase ATD designers' options.

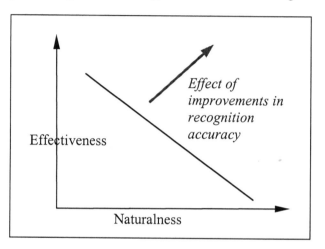

Figure 2. Relationship between effectiveness and naturalness.

In summary, task applications that produce high perplexity responses are not good candidates for current ATD technology. In such cases, for current speech recognition technology to be used, the dialogue would have to be heavily constraining—like a game of 20 questions. A certain amount of flexibility can be gained from employing word-spotting techniques. These work reasonably well for a vocabulary in the order of 40-50 words and enable the development of usable systems that mask the underlying limitations from the user. Unrestricted use of natural language remains a distant goal.

In the near-term, developers can expect technological advances that will lead to supporting larger vocabularies, improved robustness and more rapid configuration of ATDs. In turn, this will facilitate the development of more natural and usable systems, and increase the range of possible task applications.

2.2   Modality

Modality refers to the collection of physical channels across which conversational participants communicate. We are interested in modality because its specific characteristics impose constraints on what gets communicated. The effect of these constraints is that participants may perform any kind of action, but only certain actions will be conveyed by the channel. This reduction of what is conveyed may, in turn, affect the applicability of ATD technology to a given task. In this section, we will focus mainly on the limitations and implications of the telephone modality for performing ATD tasks and consider its relationship with various other modalities.

The most obvious limitations associated with the telephone modality arise from the lack of a visual channel. Humans naturally make use of physical actions such as hand gestures and head nods when communicating. Since no such visual cues are communicated in telephone-based interaction participants must adjust their communication behavior. In particular, situations where participants must compensate for the lack of a visual channel include:

♦ *Visual deixis*. Instead of "pointing" and "showing," participants must fall back on "describing" and "explaining."

♦ *Visual attention*. Studies have linked patterns of visual attention to turntaking (Argyle and Cook, 1976; Short, Williams, and Christie, 1976; Rutter, Stephenson, Ayling, and White, 1978). Instead, participants must rely on other (audible) cues to signal their turntaking intentions, such as requesting a turn and holding a turn.

♦ *Evidence of understanding*. Clark and Schaefer (1989) describe continued attention as a form of evidence of understanding. However, with the absence of a visual channel, continuing to attend is simply conveyed as silence. Thus, the speaker must rely on additional clues to judge whether the addressee heard and understood what was uttered, and whether to initiate a repair.

Another limitation of the telephone modality is that certain kinds of telephones, especially speaker-telephones and in-car telephones, only support half-duplex interaction. This means the channel can only transmit information in one direction at any one time. The net effect is that each participant is constrained either to be sending or receiving a signal, but not both. Hence, half-duplex mode removes the capability for interruptions or overlapping speech, and may distort conversants' perception of their interaction.

Further limitations of the telephone modality include the quality of the signal. Typically, telephone speech is of lower quality than equivalent face-to-face speech because of narrower bandwidth and increased channel noise. In turn, this makes speech recognition more difficult with telephone speech than face-to-face speech. Furthermore, there is a tendency for the signal quality of cellular-telephone speech to be lower still because of increased background noise—cellular telephones are often used in noisy environments such as when driving a car on the freeway.

Some modalities pick up and transmit environmental noise more than others. The issue of environmental noise is especially salient in voice-based systems. As a result, an unlikely scenario would be for all workers sharing an open office space to be using voice-based systems; the high levels of environmental noise would likely result in poor recognition accuracy.

A consideration related to bandwidth is the *resolution* of a modality. By resolution, we mean the information-carrying capacity of a modality, particularly with respect to the kinds of tokens that compose potential lexicons in the modality. Different modalities have varying limitations with respect to the kind and amount of information they can convey. For example, the keypad on a touch-tone telephone has a much smaller information capacity compared to unrestricted natural language—that is, both face-to-face or telephone speech. Also, an electronic notepad has a much smaller information capacity than a full-size workstation terminal in terms of the amount of information that can be displayed. The notion of resolution suggests that modalities can be organized compositionally with respect to their information carrying capacities. For instance, the lexicon of a keypad on a touch-tone telephone is subsumed by the capabilities of the conventional speech mode of operation. This has important implications for the applicability of tasks. In particular, anything that can be performed using the keypad of a touch-tone telephone is a natural candidate for an ATD application.

Based on these limitations, the most difficult kinds of ATD task applications to implement using telephone modality would be those that typically are carried out visually, involving the transmission of visual information (such as directions, parts, or images) or the transmission of physical action (such as assembly or dance movements).

Tasks or subtasks often have specific communication requirements and are typically performed in a particular modality (e.g., verbal-only, visual-only and text-only tasks). As a result, certain modalities may be inappropriate for particular activities. As an example, for reasons of security or privacy it may be undesirable to disclose certain information such as passwords and checking-account numbers via voice-based channels. Instead, it may be preferable to use keyboard-based entry, such as the keypad on a touch-tone telephone. Moreover, it is common to make use of several channels and for shifting among modalities to occur when carrying out tasks. Consider an everyday task like needing to speak with a colleague. One might first use a computer to ascertain whether the individual is logged-on and in the office before walking down the hallway to speak with him or her in person. This method is less intrusive and removes the potential for wasting effort if the person was not, in fact, in the office.

The ATD interface can also be considered through analogy to other more established (and conventional) modalities. For example, the U.S. Census Bureau asks people to use paper and pencil for conveying personal information such as name and date of birth. Tasks that are relatively easy to perform on paper can become quite awkward when performed verbally. As an example, consider the problem of conveying one's race via an ATD. The process of listening to 16 races listed on a written form can be tedious for the human respondent and can strain memory capacity. Developers of an ATD can use spoken-language interface heuristics (Hansen and Novick, 1996) to transform a written questionnaire into a spoken questionnaire.

Carrying the analogy farther, the kinds of human-computer interaction supported by a telephone modality could be characterized as spoken user interfaces (SUIs). SUIs may be operated in person or remotely via the telephone. One major weakness with SUIs is that it is difficult to establish a well-defined set of actions. That is, given the limitations of speech recognition technology discussed earlier, a system can encourage particular kinds of user response—by providing explicit instructions for how to respond, for example. However, it cannot constrain them in a strict sense because the user may always formulate some other response that does not comply with the explicit instructions. This lack of constraint on input is made all the more problematical by the lack of a solution to the rejection problem mentioned earlier. In contrast, a Graphical User Interface (GUI) can enforce a set of interactor objects through which the user must communicate. For example, when presenting a menu of options, only certain (typically highlighted) options may have actions defined on them. In the event that the user attempts to select one of the non-highlighted options, the system is unresponsive. In this way, the GUI can constrain the behavior of the user to a well-defined set of actions.

In a SUI, the set of interactor objects is specified by the recognition vocabulary at any given point in the dialogue. Designing spoken language dialogues that support more "natural" interaction amounts to an attempt to mask altogether the existence of these interactor objects and to create and maintain the illusion that users are permitted to express themselves in an open and unconstrained manner.

SUIs hold certain advantages over GUIs and other kinds of multi-channel interfaces. SUIs are more accessible since they can potentially be operated via the telephone, and telephones are more widely available than computer terminals. In addition, the widespread ownership of cellular telephones suggests the potential for ATD "on the move"—the GUI-equivalent of the portable computer (e.g., laptop, electronic notepad). Moreover, situations where the hands or eyes are already occupied make it inconvenient to use a GUI, whereas a SUI remains a possibility.

Conventionally, the interaction supported by a spoken-language system is audio-only, whether operated in person or remotely (i.e., via a telephone). It is possible to create multimedia modalities by incorporating several channels into a single interface. For example, it is possible to integrate a SUI with a GUI to support both audio-based and video-based interaction—i.e., spoken, graphical, textual, and video input/output (Oviatt, 1996).

2.3 Limitations Imposed By Human Conversants

So far we have considered those constraints on human-computer interaction imposed by the characteristics of the computer conversant and the modality across which interaction takes place. We now complete this picture by considering characteristics of human conversants and the limitations they bring to the design of ATD systems. These characteristics include cooperativeness, familiarity, usage, mutual knowledge, and number.

To begin, a fundamental requirement for successful human-computer interaction is that of *cooperativeness*. All ATDs, and all spoken-language systems, for that matter, with which we are familiar rely on the basic assumption that users will be cooperative, presumably because they are seeking to achieve some worthwhile goal via the system they are using. Otherwise, it is trivially easy to frustrate or confuse a spoken-language system. It is thus assumed that users will cooperate with the system in working towards achieving the overall conversational goals. More specifically, we can distinguish between two forms of cooperative behavior: 1) cooperation with respect to the task, and 2) cooperation with respect to system capabilities. In the first kind, it is expected that users remain focused and "on-task" throughout the course of the dialogue. This rules out the possibility of certain user-initiated behavior such as side sequences and tangents. Of course, the requirement that users' behavior is on-task assumes that the task is mutually known. Thus, it is important in task-based dialogues to gauge the expectations of the user and, if necessary, to establish the scope of the task explicitly—creating a shared plan for the dialogue (Grosz and Sidner, 1990).

With the second form of cooperation, it is required that the user complies with the capabilities of the ATD. This cooperation includes a wide range of behavioral aspects such as whether the system can support repairs and interruptions. This raises the issue of how users can be expected to know what the capabilities of the system are. Again, this is a problem of mutuality. There are a number of ways in which the user can form such expectations, including: 1) the system explicitly tells the user what it can and cannot do, and how the user should interact; 2) the user forms implicit expectations by entrainment (that is, the user mimics observed system behavior; if the system interrupts the user, then the user might assume that the system is capable of being interrupted itself.); and 3) the user holds certain *a priori* expectations based on past experiences, such as any prior exposure to computer systems through personal use or through accounts presented in popular media.

Closely related to cooperativeness is the *familiarity* of users both with respect to the task and interacting with the ATD. We refer to the familiarity of users in terms of levels of expertise, ranging from "naive" to "experienced" users. The degree of familiarity is a significant factor in ATD design because human expectations directly influence the nature of the interaction. At a broad level, this concerns issues like "Do users know they are talking to a computer?" This may affect the way the user phrases utterances in a manner analogous to speaking differently with an adult native speaker versus a younger or non-native speaker. More specifically, these expectations are based on issues that concern the capabilities of the system and serve to limit the style of interaction, such as "Does the system take intonation into account?" and "Can the system handle restarts and interruptions?" Users have different needs that should be addressed by the ATD. If developers intend for a system to handle a range of abilities, their effort will require careful attention to ensure that the system remains usable across the spectrum of different users.

The relationship between systems and users can also be characterized by *usage*. This refers to the extent to which the system will be used by a particular person— ranging from one-time-only to frequent usage. Usage can be regarded as orthogonal to the notion of familiarity; for instance, we may encounter experienced users— people who have a good knowledge of the specific task—who use the system infrequently. Or we may encounter naive users who use the system frequently, in which case they can be expected to acquire relevant skills with time. Thus, with usage we are primarily concerned with the opportunity for users to learn how to interact with the system and to become proficient with performing the task.

Both usage and familiarity are relevant when considering the learning curve of an ATD. Familiarity determines the starting point on the curve and usage determines the potential for advancing along the curve. The learning curve associated with a particular task application, along with factors such as convenience ("Is it convenient to use?") and effectiveness ("Is it effective for achieving the task?"), all contribute towards the eventual acceptance or failure of a system.

Another important factor that influences the suitability of a task is *mutuality* of reference. Mutuality of reference describes the condition in which conversants believe they are referring to the same thing. In general, there may be many different ways of expressing an idea and referring to a specific concept or object. Accordingly, referring expressions presuppose that the other conversant will have appropriate knowledge to identify the referent. Moreover, effective task dialogue depends on the conversants being able to accurately gauge the capabilities of their collaborators to make use of the resources and to know what knowledge is shared. From an ATD standpoint, mutuality of reference is easiest to establish when there are canonical means of expression. For example, it is generally understandable when a banking ATD refers to "your account" because customers normally know that this term refers to their checking account. This reduces the variation and uncertainty in anticipating how the user may refer to something specific. Some of the differences in expression may result from the fact that the user may not know exactly what they want; that is, they have a problem they want solved but do not already have the solution in mind. A more basic factor underlying this problem is that the user may know what their need is but not be aware of what the system knows. This lack of *a priori* mutuality typically arises because systems generally do not belong to known communities, so the user has little or no basis for inferring indirect co-presence of referents (Clark and Marshall, 1981).

Finally, humans are capable and proficient at interacting both with other individuals and with groups of people. Current ATD technology, however, is largely aimed at supporting only dyadic interaction. There are many interesting applications involving multi-party interaction, such as capturing spontaneous voice-based collaboration for later retrieval (Hindus and Schmandt, 1992). However, there are a number of open issues associated with supporting multi-party interaction, such as identifying who is speaking and to whom, that need to be addressed (Novick and Walpole, 1990; Novick, Walton, and Ward, 1996).

3.   CHARACTERIZATIONS OF TASK

Our discussion so far has focused mainly on identifying constraints on human-computer interaction imposed by characteristics of the conversants—human and computer—and the modality supporting the interaction. These limitations directly affect the applicability of both ATD and non-ATD tasks. We now proceed to consider a selection of general characteristics of tasks that provide further insight into determining the suitability of a task for development as an ATD. In particular, in the following sections we will describe some complementary perspectives on tasks that provide an analytical basis for making such judgments: 1) communications requirements; and 2) group action.

3.1   *Communication Requirements Perspective*

A useful analytical view of task is that of the flow of information between the conversants. By information flow we mean task-relevant information that is communicated by way of assertions. There are two dimensions of interest: amount and direction. The first dimension is the amount of information communicated—measured on a per-task rather than a per-utterance basis. As a metric, the amount of information conveyed can be characterized in terms of low, medium or high levels of flow. The second dimension of interest is the direction of information flow. From a system standpoint, we can distinguish between information input (information flowing from the user to the system) and information output (information flowing from the system to the user). Table 1 illustrates the relative information flow of different kinds of tasks with respect to these dimensions of level and direction.

Table 1.   Information flow of various tasks.

Information Flow	Low Output	Medium Output	High Output
Low Input	Telephone polling	Ski/weather report	Credit history
Medium Input	Pizza order	Vehicle navigation	On-line encyclopedia
High Input	Questionnaire	Auto mechanic	Advisory Service

Information flow is a useful perspective because of the relative burden it places on specific components of an ATD. In particular, high levels of information input make it harder for speech recognition whereas high levels of information output make it harder for the dialogue model. Given current ATD technology, we regard the most challenging tasks to be those that entail high levels of information input because of the problems outlined in Section 2.1. High levels of information output are less of a problem to the system, although the consequences of outputting inappropriate information can be costly to the user. Furthermore, misrecognition might be the cause of inappropriate output.

Consequently, the most effective ATD applications are simple tasks involving low input and output of information. Such task characteristics minimize the potential for misrecognizing user input and for producing inappropriate system output. However, many of the more "interesting" tasks entail higher levels of information flow (both in terms of input and output). We will now briefly consider how we can support some of these larger and more interesting tasks.

Two important strategies for supporting tasks with higher levels of information flow involve: 1) task selection, and 2) dialogue control. Even within task characterizations based on information flow, some of the high input and high output tasks are better suited as ATD applications than others. The first strategy, task selection, entails taking into consideration some other factors such as domain semantics, for example. Limited domain semantics is an important requirement for more accurate prediction. The overall size of the domain is not the main issue; rather, it is the breadth of the domain that is the most significant factor. The breadth of a domain can be measured in terms of branching factor. This is especially true with respect to reference; if there were a way of referring uniquely to large numbers of items, and if the user knew the appropriate referential keys, then the size of the domain is less of an issue. For example, consider a catalogue sales operation in which the user has the catalogue at home while they call. In this case, they can know an alphanumeric reference for each item in the catalogue and can use the reference to obtain information about items or to order items. So, in general, wide flat domains (like supporting access to the Yellow Pages) are poor choices for ATD applications, while narrower, perhaps deeper domains would be better.

Another related consideration is task planning. Performing a task entails the notion of a plan that maps out the nature of the interaction in terms of the *goals* (what you want to achieve) and *actions* (the steps involved in achieving a goal). The plan gives rise to a set of expectations of the role to be fulfilled by the other conversant. More specifically, there are six levels at which plans are relevant to ATD interaction. These levels can be expressed in terms of the system and user as follows:

♦   System's plan,

♦   System's beliefs about user's plan,

♦   System's beliefs about user's beliefs about system's plan

♦   User's plan,

♦   User's beliefs about system's plan, and

♦   User's beliefs about system's beliefs about user's plan.

For effective plan execution, the system and user must share the same plan or at least have compatible plans. For example, the system's plan may be to conduct a survey in which it asks the user questions in order to establish certain information. The user, on the other hand, may have a plan to cooperate with the system but may not know in advance precisely what information will be required. In this sense, the system's plan and user's plan are not the same but are compatible. From a system standpoint, we can identify three task variants: 1) fixed tasks, 2) template tasks, and 3) open-ended tasks. Fixed tasks are typically well defined and completely known in advance, for example, calling to register a vote by phone. Template tasks involve a stereotypical set of events although specific information typically is missing; these can be conveniently represented in terms of templates with "information slots" that need to be filled, for example, when ordering a pizza. Finally, with open-ended tasks the underlying purpose may not become known until the dialogue is under way. For example, with medical diagnosis, it is not known what is wrong with the person in advance of the dialogue and so it is difficult to anticipate what actions will be taken until the dialogue unfolds.

Of these task variants, fixed tasks are the easiest to support using ATD technology, and template plans are mostly doable given limited domain semantics for the missing information. Open-ended tasks, however, are especially problematic because it is not well understood how to construct task plans dynamically.

The second approach to handling higher-volume tasks involves maintaining control over the flow of information. In section 2.1 we discussed various methods for dialogue control, namely, using entrainment, explicit constraints, and conversational initiative to shape the user input. As an example, for the purpose of making more accurate predictions as to what the user will utter next, it is preferable to induce an assertion by requesting specific information, as opposed to waiting for the user to volunteer the information, especially when there is a great deal of information to be gathered. This is what was attempted with the census task (Cole et al., 1997).

3.2   Group Action Perspective

The goal of this section is not to provide a comprehensive survey of sociological views of tasks but to give a sampling of social science perspectives on how tasks might be characterized. From these characterizations, we hope to gain insights into the kinds of tasks for which spoken-language technology might be appropriate. Our discussion grows out of a review of task classification by McGrath (1984).

Several efforts have been made to characterize group tasks. McGrath and Altman (1966) identified six factors for classifying tasks: 1) properties or dimensions of task *qua* task; 2) behaviors required by task; 3) behaviors usually elicited by task; 4) relations among behaviors of group members; 5) goal of task and product of task; and 6) criterion of task. Of these, factors 1, 5, and 6 relate to the task to be performed, while factors 2 and 3 describe the relationship between the task and the task performer. Factor 4 describes the relationship between task participants.

Another model of group tasks is that of Shaw (1973). According to Shaw's model, there are four key factors: 1) intellective vs. manipulative; 2) relation between task and doers in terms of a) difficulty, b) intrinsic interest, and c) population familiarity; 3) solution (multiple vs. specific); and 4) cooperation. Here, factors 1 and 3 are useful for characterizing the nature of a task and its solution. Factor 2 describes the relationship between the task and the performers of the task, while factor 4 again describes the relationship between task participants.

The factors enumerated by McGrath, Altman and Shaw are largely suitable for direct application as features in the feature space described in Section 4. However, while these models are useful for characterizing many kinds of social science tasks (or experiments), there are serious questions as to their coverage and exclusivity.

McGrath subsequently proposed an analytic scheme for group tasks that integrated earlier models of task with the goal of characterizing group tasks into categories that are mutually exclusive, collectively exhaustive, and useful. Within McGrath's model, four general processes (generation of alternatives, selection from alternatives, negotiation and execution) are placed within two broad dimensions of conflictive vs. cooperative and conceptual vs. behavioral tasks (see Figure 3). In all, McGrath's model (or "circumplex") classifies group tasks into eight categories: planning tasks, creativity tasks, intellective tasks, decision-making tasks, mixed-motive tasks, cognitive conflict tasks, contests (or battles), and performances. We will briefly evaluate each of these task types in terms of their amenability to being supported by an ATD.

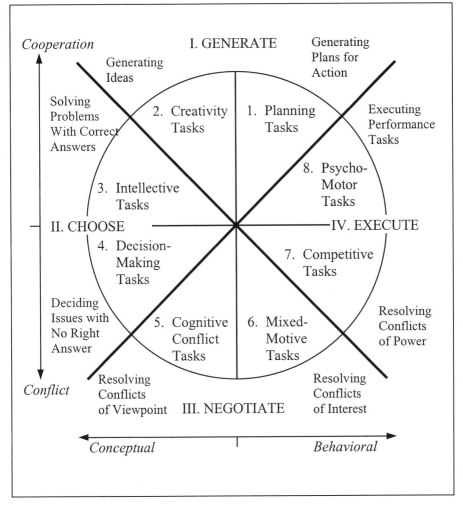

Figure 3. McGrath task circumplex.

Planning tasks (task type 1) are those where participants are collaborating in generating plans for action. As such, they are characterized as being more behavioral than conceptual, and more collaborative than conflictive. The element of cooperation is an important factor for ATD treatment. Furthermore, a computational agent may be a good partner for a planning effort, given the ability to quickly follow the implications of a particular plan element within an established plan template. On the other hand, plans that revolve around new information may put strains upon the ability of designers to represent the needed domain knowledge. Additionally, typical planning tasks may be characterized by a strong orientation toward mixed-initiative. This kind of free-form give and take doesn't lend itself to the kinds of techniques (especially control over the conversational initiative) used to mitigate speech recognition limitations.

The second task type, the creativity task type, shares the generative and cooperative aspects of task type 1, but is characterized as being more conceptual than behavioral. Oddly, these kinds of tasks may be good candidates for ATD treatment, since there may be less of a requirement that the computational agent be authoritative and directly relevant in its responses. A typical task of this type is brainstorming, an area where the computational agent can cooperate by offering various perspectives or techniques and letting the human put the pieces together in novel ways. This separation of roles may be more amenable to a single-initiative style of interaction than is the case in tasks where participants share their roles.

The third task type, that of intellective tasks, is also characterized as being cooperative and conceptual, but differs from task type 2 in its orientation toward choosing answers instead of generating ideas. Tasks of this type are distinguished by there being a correct answer, either a logically compelling one, or one arrived at by consensus of experts. These kinds of tasks may be the most amenable of all of McGrath's task types for implementation as an ATD. Indeed, it is easy to see that tasks such as entering census data and getting directions for navigating a car both fit within this type. Although, in general, tasks of this type are not performed, necessarily, with a single-initiative style of interaction, they often translate well into such a style.

Decision-making tasks (task type 4) differ from intellective (type 3) ones in the degree of conflict involved. The key notion for this task type is that there is a preferred answer, or one that is agreed upon, rather than one that is correct *a priori*. Given the nature of the way answers are arrived at within this task type (i.e., by consensus), these kinds of tasks may not be well suited to ATDs. The consensual nature of these tasks is predicated on shared values or morals that might be less than compelling when coming from what is known to be a computational agent.

Task types 5 and 6 make up the "Negotiation" quadrant of the task circumplex and are characterized by their orientation toward resolving conflicts, either of viewpoint (task type 5: cognitive conflict tasks) or of motive (task type 6: mixed-motive tasks). While the notion of convincing or persuading a computational agent is, at best, problematical, when the persuasion is going the other direction, the prospect seems less unlikely. Still, the expected asymmetry of the flow of negotiation would lead us to expect that ATDs implementing these kinds of tasks would be unpalatable to human conversants unless care is taken to mitigate the negotiative imbalance.

Type 7 tasks (contests and battles) are characterized as being conflictive (of course) and behavioral (as opposed to conceptual). One need only look at the success of video games to appreciate the applicability of computational realization to these kinds of tasks. The only question is the extent to which these kinds of tasks can be carried out verbally, and the extent to which these verbal tasks can be realized by an ATD. Here it is worthwhile to emphasize the distinction between what might be called "task cooperation" and "interaction cooperation." While task cooperation may not be necessary for ATD-human interaction, interaction cooperation is essential. Although it is possible to distinguish these forms of cooperation theoretically, it is unclear whether it is always possible to separate the two in practice.

The final task type (performance) resides within the same cooperative and behavioral sector of the circumplex as task type 1 (planning task type), differing in its orientation by its emphasis on psycho-motor activity. While it is possible to imagine ATD tasks that contain psychomotor elements (an on-line engine diagnosis and repair service, for instance), the emphasis on physical action makes these kinds of tasks unlikely for delivery via ATDs.

These eight task types, then, as characterized by the four quadrants and two dimensions, are meant to provide a mutually exclusive, exhaustive and useful classification of tasks. Although none of the task types is completely antithetical to ATD technology, clearly some (types 2 and 3 especially) are conducive to more straightforward implementation. The other task types could be candidates for ATD treatment if the objections raised above are adequately addressed.

Although straightforward and initially appealing, McGrath's model has a number of problems when applied to ATD task analysis. These difficulties arise from the purposes of the model and from its inability to account for certain classes of observed behaviors.

In developing the circumplex model of task, McGrath was aiming at describing tasks used by psychologists in experimental situations. These tasks tend to be short-term, often artificial, and certainly not *situated* in the workaday world (cf., Moran and Anderson, 1990). In contrast, the point of a task model for ATD design is to identify tasks that are situated (cf., Suchman, 1987), and that reflect users' real needs. Another problem of the model's purpose is that it was designed for group tasks rather than dyadic ones. We expect that telephone-based spoken language systems of the foreseeable future will be most often designed for interaction between a computer and a single user.

Our own experience in applying McGrath's model to situated group tasks suggests that the model does not account for certain aspects of tasks, including kinds of cross-task activity and perhaps some additional task types. These additional task types are particularly important for ATD purposes because they relate most directly to communication. We conducted a systematic coding of tasks in interactions videotaped in actual work situations. Based on our analyses, we propose that McGrath's eight task types be augmented by a) information gathering and b) information sharing. Across all task types, we also find an additional dimension that describes the relationship of the current activities to the underlying domain task. This dimension categorized the group's activities as being on-task, engaged in sub-goal digression, engaged in a meta-task, or orthogonal to the underlying task.

A more perplexing problem with these sorts of task classification schemes is that in real-world collaborative tasks there are often several different task types simultaneously at play. For example, within a specific human-human interaction, conversants might be resolving mixed motives, and competing for power within the context of brainstorming. It is not clear whether a particular action (or, in the case of verbal tasks, a particular utterance) can be uniquely associated with a single task type.

While McGrath's model does not explicitly mention task decomposition, we have, in our own use of the model, examined human-human conversations in terms of tasks and sub-tasks, using the model on larger and smaller units of conversation.

In general, it is possible to view fixed tasks as hierarchically decomposable tasks made up of sub-tasks, which may themselves be further decomposed until a "primitive" level is reached.

4.   AN ATD FEATURE SPACE

Section 2 described some of the constraints on interaction posed by computers, modalities, and humans. Section 3 presented a sample of perspectives that characterized tasks. In this section, we propose to integrate these views. In the abstract, the constraints imposed by the user, system, modality and task form a complex feature space, where states in the space correspond to task characteristics that are well- or ill-served by different constraints. Such a feature space would be useful because it would indicate the suitability of newly considered tasks for ATD delivery. Although a complete account of the feature space is elusive, we can outline the main dimensions that define it. The applicability of an ATD approach to a particular task varies with the characteristics of the user, the system, the modality through which they communicate, and the task itself. As summarized in tables 2 to 11, our analysis identified 42 features in this space. With each feature, we have associated a symbol representing the effect of the feature on the suitability of creating an ATD system. A "+" indicates that the feature is positively associated with the success of an ATD. A "-" indicates that the feature is negatively associated with the success of an ATD. And a "*" indicates that effects of the feature on the success of an ATD depend on the particular circumstances or on details internal to the feature.

Features 1 to 3 relate to the system dimension only. Positive features relating to the system include the recognizer's accuracy rate and robustness, including ability of system to handle out-of-vocabulary words. Negative features relating to the system include the length of time taken to process a user input and provide an appropriate system response.

Feature 4 relates to the user dimension only. Positive features relating to the user include the expertise of the target user population generally. That is, it is probably better to develop a system aimed at users who are already familiar with similar tasks or interfaces.

Features 5 to 7 relate to the modality dimension only. Positive features relating to the modality include the quality of the channels making up a modality, including insensitivity to noise and the channel's bandwidth and resolution. Negative features relating to the modality include limitations of the modality supporting the interaction, such as lack of ability to refer. Features relating only to modality whose effects may vary depending on the particular circumstances include the availability of alternate or supplementary modalities (i.e., because referring becomes easier).

Features 8 to 14 relate to the task dimension only. Positive features relating to task include the extent to which the task is executable through speech acts that are appropriate to the partners in the dialogue. For example, it is reasonable to expect an ATD to perform *inform* acts with respect to domain information, but it is less reasonable for an ATD to perform acts that are not suited to a machine, such as expressing sympathy. Negative features relating to task include task size, the breadth of domain semantics, the size of the recognition vocabulary, the perplexity of expected user responses, and the confusability of words in the recognition vocabulary. Features relating to task whose effects may vary depending on the particular circumstances include the characterization of tasks in terms of task types; developers should consider the effects of task type on ATD suitability as discussed in Section 3.

Features 15 to 23 relate to the system and the user together. Positive features relating to system x user include: 1) the use of explicit constraints in controlling user input; 2) reliance on entrainment to influence user input; 3) the ability of the system to handle meta-dialogue (such as repairs and help); 4) the extent to which the interaction feels intuitive and natural to the user; 5) the expertise of the target user population with respect to the system 6) familiarity of the system capabilities; 7) the level of cooperation by the user with respect to system capabilities; and 8) the degree to which the system may adapt to the user. Features relating to system x user whose effects may vary depending on the particular circumstances include the varieties of initiative supported in the interaction.

Feature 24 relates to the system and the modality together. Positive features relating to system x modality include the ability of the system to shift modalities.

Features 25 and 26 relate to the system and the task together. Positive features relating to system x task include the degree to which the system's run-time interaction goals can be made explicit at design time, and the degree to which the task is representable in the system's underlying formalism.

Features 27 and 28 relate to the modality and the user together. Positive features relating to modality x user include the ability of the user to shift channels. Negative features relating to modality x user include the level of noise in the user's environment and the sensitivity of the modality to noise.

Features 29 to 32 relate to the modality and the task together. Positive features relating to modality x task include the extent to which the modality supports sharing of resources needed to accomplish the tasks and the extent to which particular task components relate directly to the nature of the modality. Negative features relating to modality x task include the extent to which additional resources are needed to perform sub-tasks and the quantity of information flowing between the user and the system.

Features 33 to 39 relate to the user and the task together. Positive features relating to user x task include: 1) the degree to which the user's goals in this task are predictable; 2) the expertise of the target user population with respect to the task; 3) the degree to which the task is stereotypical; 4) the degree to which users may become adept at performing the task; 5) the degree to which users are already familiar with the task; and 5) the level of cooperation by the user with respect to the task. Negative features relating to user x task include the steepness of the learning curve associated with performing the task.

Features 40 to 42 relate to the user, the system and the task together. Positive features relating to user x system x task include the degree to which the task plan is shared by the system and user, the expected frequency of usage of system by the user for the task, and the usability of the system for the task.

Table 2.   System features.

No.	Effect	Feature
1	+	Recognition accuracy
2	-	System response time
3	+	Robustness

Table 3.   User features.

No.	Effect	Feature
4	+	User expertise generally

Table 4.   Modality features.

No.	Effect	Feature
5	+	Channel quality
b		Insensitivity to noise
a		Bandwidth
6	*	Availability of other modalities
7	-	Modality interaction limitations

Table 5.   Task features.

No.	Effect	Feature
8	-	Task size
9	-	Breadth of domain semantics
10	-	Vocabulary size
11	*	Task types
12	+	Suitable speech acts
13	-	Perplexity
14	-	Confusability

Table 6.   System x user features.

No.	Effect	Feature
15	+	Explicit constraint of user input
16	+	Entrainment of user input
17	*	Initiative
18	+	Meta-dialogue abilities
19	+	Intuitive feel
20	+	User expertise for system use
21	+	User familiarity of system capabilities
22	+	User cooperation
23	+	System adaptation to user

Table 7.   System x modality features.

No.	Effect	Feature
24	+	System can shift channels

Table 8.   System x task features.

No.	Effect	Feature
25	+	Predictable system goals
26	+	Representable task

Table 9.   Modality x user features.

No.	Effect	Feature
27	+	User can shift channels
28	-	Sensitivity to noisy environment

Table 10. Modality x task features.

No.	Effect	Feature
29	+	Modality supports shared resources
30	+	Modality supports task components
31	-	Resources needed for sub-tasks
32	-	Quantity of information

Table 11. User x task features.

No.	Effect	Feature
33	+	Predictable user goals
34	+	User expertise for task
35	+	Stereotypical task
36	-	High task learning curve
37	+	Users can become adept
38	+	Familiarity of the task
39	+	User cooperative for task

Table 12. User x system x task features.

No.	Effect	Feature
40	+	Shared plan
41	+	Frequent task
42	*	Speaker-dependence

In using these results, two limitations of our analysis should be considered. First, our analysis did not identify features that involved user *x* system *x* modality, user *x* task *x* modality, system *x* task *x* modality, or user *x* system *x* task *x* modality. However, the fact that our survey did not disclose such features does not mean that they cannot or do not exist. Second, the analysis does not provide a means of comparing strength of features. Our analysis only looked at direction, not extent. Thus in considering use of ATD technology, it would be best to consider the features qualitatively rather than simply adding up the numbers of plusses and minuses. Some of the 42 features have elements in common or appear otherwise related. Other features, and groups of features, seem to be entirely orthogonal. Future work in refining the feature space may focus on grouping and abstracting the features.

5. CONCLUSION

The range of tasks suitable for development as a spoken language system can extend far beyond dialogue-based access to information. The components of the typical ATD task—collecting and providing information—can be arranged and rearranged to encompass a variety of interactive activities. In point of fact, tasks for ATDs often have limited dialogue requirements. Frequently these tasks involve single queries, such as in a yellow-pages or other directory service. In such cases, there is little scope for complex exchanges. More complex dialogues might occur in advice services, such as automobile repair or medical information. In these domains, analogous human-operated services already exist. For example, in the domain of medical advice the user calls with a problem, and (typically) a nurse offers advice over the phone. The human advice system shares with a possible ATD in this domain the limitations of the modality. In-person diagnosis typically involves the medical professional examining the patient visually or physically; obviously this is not possible over the telephone.

By rearranging the quantities and direction of information flow, one can develop other classes of tasks for ATDs, such as dialogue-based *collection* of information (DBCI). ATD applications within DBCI are characterized by elicitation of information from users; the tasks include, for example, questionnaires for marketing, forms for warranty registration or service requests, questionnaires for the U.S. census (see Cole et al., 1997), and even templates for routing telephone calls in an organization.

More generally, choice of modality reflects certain economies of human-human interaction. In general, reading is harder than hearing; writing is harder than speaking; and speaking writing (i.e., reading out loud) is harder than reading. Other things being equal, spoken-language modalities would appear to have natural advantages. However, this broad perspective may not hold when we take task into account. For certain tasks, writing, drawing or pointing may be preferred over speaking either because of requirements of the task itself or because of the context in which the task is performed. But it would be a mistake to assume that tasks always require "traditional" methods and modalities in order to be accomplished effectively. The technology of interaction has an extensive history of shifts in modality that supplant traditional means for accomplishing known tasks. These relative successes include telegrams, fax, voice messaging, electronic mail, telephone menus, and electronic transactions. In each case there is a learning curve and a winnowing out of methods that are ineffective; of course, many other putatively innovative technologies failed for lack of a comparative advantage. The features enumerated in Section 4 are intended to provide a more systematic approach to picking the winners. In developing ATD applications over the short term, we seek to take advantage of the flexibility and adaptability of human behavior in overcoming the limitations of ATD technology. In the medium to long term we look forward to technological advances that will improve the applicability of ATD systems and thus extend the range of tasks for which ATDs can exploit the advantage of their natural ease for human interaction.

ACKNOWLEDGMENTS

This work was partially supported by US WEST and the members of the Oregon Graduate Institute's Center for Spoken-Language Understanding.

REFERENCES

Argyle, M., and Cook, M. (1976). *Gaze and mutual gaze*. Cambridge, UK: Cambridge University Press.

Brennan, S. E. (1991). Conversation with and through computers. *User Modeling and User-Adapted Interaction, 1*, 67-86.

Clark, H. H., and Marshall, C. R. (1981). Definite reference and mutual knowledge. In: A. K. Joshi, B. Webber, and I. A. Sag (Eds.), *Elements of discourse understanding* (pp. 10-63). Cambridge: Cambridge University Press.

Clark, H. H., and Schaefer, E. F. (1989). Contributing to discourse. *Cognitive Science, 13*, 259-294.

Clark, H. H., and Wilkes-Gibbs, D. (1986). Referring as a collaborative process. *Cognition, 22*, 1-39.

Cole, R., Hirschman, L., Atlas, L., Beckman, M., Biermann, A., Bush, M., Clements, M., Cohen, J., Garcia, O., Hanson, B., Hermansky, H., Levinson, S., McKeown, K., Morgan, N., Novick, D., Ostendorf, M., Oviatt, S., Price, P., Silverman, H., Spitz, J., Waibel, A., Weinstein, C., Zahorian, S., and Zue, V. (1995). The challenge of spoken language systems: Research directions for the Nineties. *IEEE Transactions on Speech and Audio Processing, 3*(1), 1-20.

Cole, R., Novick, D. G., Vermeulen, P. J. E., Sutton, S., Fanty, M., Wessels, L., de Villiers, J., Schalkwyk, J., Hansen, B., and Burnett, D. (1997). Experiments with a spoken dialogue system for taking the US census. *Speech Communication, 23*(3), 243-260.

Grosz, B.J., and Sidner, C.L. (1990). Plans for discourse. In P. Cohen, N. Morgan, and M. Pollack (Eds.), *Intentions and communication* (pp. 417-444). Cambridge, MA: MIT Press.

Hansen, B., and Novick, D. G. (1996). Systematic design of spoken prompts, *Proceedings of the Conference on Human Factors in Computing Systems (CHI '96)* (pp. 157-164). New York: ACM.

Hindus, D., and Schmandt, C. (1992). Ubiquitous audio: Capturing spontaneous collaboration. *Proceeding of the ACM 1992 Conference on Computer-Supported Cooperative Work (CSCW '92)* (pp. 210-217). New York: ACM.

Kennedy, A., Wilkes, A., Elder, L., and Murray, W. (1988). Dialogue with machines. *Cognition, 30,* 37-72.

Kitano, H., and Van Ess-Dykema, C. (1991). Toward a plan-based understanding model for mixed-initiative dialogues. *Proceedings of the 29th Annual Meeting of the Association for Computational Linguistics* (pp. 25-32). Somerset, NJ: ACL.

McGrath, J. E. (1984). *Groups: Interaction and performance.* Englewood Cliffs, NJ: Prentice-Hall.

McGrath, J. E., and Altman, I. (1966). *Small group research: A synthesis and critique of the field.* New York: Holt, Rinehart & Winston.

Moran, T. P., and Anderson, R. J. (1990). The workaday world as a paradigm for CSCW design. *Proceedings of the Conference on Computer Supported Cooperative Work* (pp. 381-393). New York: ACM.

Novick, D. G., and Walpole, J. (1990). Enhancing the efficiency of multiparty communication through computer mediation. *Interacting with Computers, 2*(2), 227-246.

Novick, D. G., Walton, L., and Ward, K. (1996). Contribution graphs in multiparty conversations, *Proceedings of the International Symposium on Spoken Dialogue (ISSD-96)* (pp. 53-56). Tokyo: Accoustical Society of Japan.

Oviatt, S. L. (1996). Multimodal interfaces for dynamic interactive maps. *Proceedings of the Conference on Human Factors in Computing Systems (CHI '96)* (pp. 95-102), New York: ACM.

Rutter, D., Stephenson, G., Ayling, K., and White, P. (1978). The timing of looks in dyadic communication. *British Journal of Social Clinical Psychology, 17,* 17-21.

Shaw, M. E. (1973). Scaling group tasks: A method for dimensional analysis. *JSAS Catalog of Selected Documents in Psychology, 3,* 8.

Short, J., Williams, E., and Christie, B. (1976). *The social psychology of telecommunications.* London: John Wiley.

Smith, R. (1994, Feb.). Spoken variable-initiative dialogue: An adaptable natural-language interface. *IEEE Expert,* 45-50.

Suchman, L. A. (1987). *Plans and situated actions.* Cambridge: Cambridge University Press.

Walker, M., and Whittaker, S. (1990). Mixed initiative in dialogue: An investigation into discourse segmentation. *Proceedings of the 28th Annual Meeting of the Association for Computational Linguistics* (pp. 70-78). Somerset, NJ: ACL.

Zoltan-Ford, E. (1991). How to get people to say and type what computers can understand. *Journal of Man-Machine Studies, 34,* 527-547.

9 WHY DO PEOPLE DIAL WRONG NUMBERS?

Arnold M. Lund

U S WEST Advanced Technologies

Key words: Dialing Errors, Dialing Speed, Touch-Tone, Interactive Voice Response (IVR)

Abstract: *What is the lowest error rate that might reasonably be expected when people use an interactive voice response (IVR) system? What are some of the sources of errors during touch-tone entry? This paper addresses the most fundamental aspect of an IVR dialogue, the errors arising from the use of the touch-tone pad itself. Data are reported that provide baseline error rates and digit entry performance measures. A model of the process of entering a simple string is derived, and it fits the data well. This model suggests where errors might arise during more complex IVR dialogues and parameters that will be useful in predicting performance when comparing IVR dialogue designs.*

1. INTRODUCTION

A telecommunications company marketing person asked whether it would be possible to guarantee that customers would make no errors on new telephone services. This would be an important performance requirement if it was possible and would be a goal consistent with the argument that human error is largely a function of the system design. It would be important because the touch-tone telephone is the most widely deployed interface device for controlling computer applications. In the U.S., there is at least one telephone in over 94% of homes and in many cases more than one. Roughly 90% of these are touch-tone phones (estimates range from 81% to 96%), and the penetration of touch-tone phones is increasing. The touch-tone telephone is used not only for establishing connections with others over the computer-based telephone network, it is used to access information services and to control applications such as messaging and banking. New telephones are even emerging that are small computers, with the touch-tone pad as a primary method of input. The pad is used in security interfaces and other applications requiring accurate data entry. The touch-tone button layout and telephony metaphor are even being integrated into workstation software, and are being used for interactive television controls.

What is the lowest error rate that might reasonably be expected during touch-tone dialogues, and what approaches might be used to drive application performance closer to this goal? The simplest task people perform with the touch-tone pad is to dial a telephone number to place a call. The touch-tone telephone has been widely studied and improved since the earliest human factors design research was conducted (e.g., Deininger, 1960; Wikell, 1982), and most people have had years of experience with it. People should be well past the rapidly changing part of the learning curve, having achieved major gains in speed (according to the power law of practice). Near asymptotic accuracy might be expected. More complex uses of touch-tone dialing such as interactive voice response (IVR) applications can only be expected to increase errors or slow performance relative to the baseline of the simple, well-practiced dialing task. However, the speed and accuracy averages observed in the simple dialing task should be useful for GOMS analyses (Card, Moran, and Newell, 1983) and similar techniques for predicting and improving the usability of these more complex designs.

Several years ago, Chapanis and Moulden (1990) noted that they were unable to find any research directly identifying which numbers were easiest to remember and only a few studies with hints about the memorability of specific numbers. There is a similar shortage of work focussed specifically on the accuracy of typing numbers using telephone touch-tone keypads, where the focus of interest is on the task of number entry itself. As the telephone industry and public utility commissions are considering whether to mandate longer telephone numbers for the hundreds of millions of phone calls that are placed every day, or whether to further divide area codes, designers in the industry have been troubled by the lack of information that would allow a prediction of the impact of the errors on network demand caused by redialing and on the public receiving wrong numbers. There is even less research on novel dialing tasks that have been emerging such as those based on entering text strings using a touch-tone pad, and no research comparing strings with common elements such as 1-800 with those of comparable length but without the common element.

While there is a shortage of work specifically focussing on dialing errors, there have been several studies using dialing error rates as a dependent variable. Research using error rates during number entry as the dependent variable generally addresses specific design decisions such as the optimal size of a key or arrangement of numbers on a key pad. For example, Copping (1974) reviewed some of the previous studies that report error rates and number entry time and reported an experiment manipulating the length of number strings. The focus of the study, however, was the comparison of rotary and touch-tone dialing. With that focus, the types of errors made and the possible cognitive and motor processes leading to those errors were not discussed. There was no attempt in the study to define a baseline of number entry performance.

While the authors of these earlier studies were not interested in number entry performance per se, their work can be used to obtain a preliminary view of how performance might vary as a function of number length. Despite differences in purpose and methodological details, many of the studies do include one or more conditions in which the subjects used a touch-tone telephone (with the traditional 3 x 4 format pad illustrated in Figure 1 (although sometimes without the * and # sign) to enter a number that was continuously present during the dialing task. Over 30 studies were uncovered. The studies extend from 1985 back to 1958, with the majority of the studies being conducted in the 1960s and early 1970s. Obviously one of the facts that must be kept in mind when interpreting the data is that many of the earliest studies were designed to test requirements for the touch-tone pad itself, and therefore the subjects often had never experienced touch-tone dialing before.

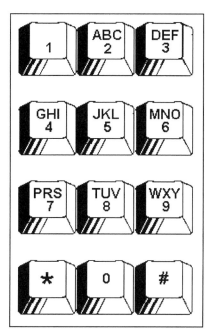

Figure 1. A typical 3 x 4 key layout.

A meta-analysis can be conducted by plotting the touch-tone entry conditions together in a common graph. Figures 2 and 3 show all the conditions in which the subjects used touch-tone pads to enter numbers of various lengths while the numbers were present during dialing. Figure 2 shows how the likelihood of entering a string in error (i.e., where one or more of the digits in the string was entered in error) varies as a function of number length. Figure 3 shows how the total time to enter the string (defined here as the time between entering the first digit in the number and the last digit in the number) varies by number length. Error and timing data were not both available for every condition, so the figures do differ in the number of conditions plotted.

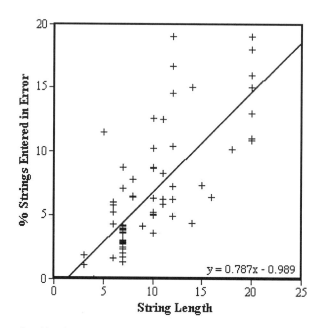

Figure 2. Number entry accuracy as a function of digit string length.

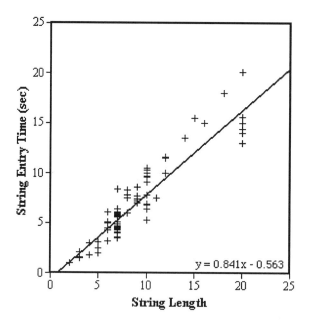

Figure 3. Number entry speed as a function of digit string length.

A line fitted to the 75 conditions included in Figure 2 accounts for 58% of the variance. This is consistent with an analysis conducted several years ago using proprietary studies and plotting 22 conditions that accounted for 56% of the variance. That study had been designed to help answer questions about the appropriate length of security codes entered using a touch-tone pad. The variance accounted for by a line fitted to the 94 conditions in Figure 3 was 86%. It should be noted that some of the articles only showed errors in graphs, or reported them in terms of the probability of entering a specific digit in error, and thus variance was introduced into the data that was not present in the original studies. Further, there was occasionally ambiguity concerning exactly how the string entry time was calculated, and so again it is possible that variance was introduced as a result.

Despite the variety of goals, equipment, specific stimuli and details of presentation, and other experimental conditions, trends in the data are apparent. In general, the percentage of the number strings entered with at least one digit in error increases as the string length increases. For the 7- and 11-digit telephone numbers that are entered in the millions every day, more than one out of every twenty numbers is entered in error. For the longer strings such as credit card numbers, one out of every ten or more attempts is in error. This suggests that the error rates associated with simple dialing may be considerably higher than is often assumed, and the social cost of errors during activities such as credit card entry could be higher than is typically realized.

The number entry times seem to be well described by a linear model. The inter-digit entry time estimate is 841 msec. When the linear relationship in the timing data was first observed in the earlier study using proprietary data, it was surprising. The expectation had been that the inter-digit times would be much faster for dialing strings that could be completely rehearsed in working memory than for longer strings (where the subject would have to look back at the string being entered or where the subject might need to adopt a strategy to deal with memory failure due to interference). As a result, we were predicting a shallower number entry time slope for numbers seven digits or less than for those ten digits and greater. There is a suggestion of such an effect in Figure 3, but there is sufficient noise in the graph to make it suspect.

It is possible, of course, that subjects either use chunking strategies to keep even the longer 20-digit numbers in working memory, or they may keep their inter-digit entry times constant and in making a speed-accuracy tradeoff shift the impact of longer numbers to the errors they make. Unfortunately, the variance in Figure 2 is sufficiently great that it is not possible to have a high degree of confidence about whether the underlying relationship between errors and number length is linear or curvilinear. A broader range of number lengths, including the numbers with 30 or more digits that are becoming common as part of electronic commerce and other interactive applications, might reveal more effectively the underlying relationship. As a result, it was decided to conduct a more controlled study of errors and entry time as a function of string length and type. In addition, given that it had been 25 years since many of the studies in the meta-analysis were conducted, and that touch-tone is essentially ubiquitous and has been for many years, an updated baseline of the touch-tone performance of the now expert dialers was needed.

2.   METHOD

2.1. Subjects

Fifty-two subjects representing a broad range of telephone users were recruited for the experiment by an agency for hiring temporary help. Twelve of the subjects were dropped from the experiment because of incomplete data and other factors, leaving 40 subjects (12 male and 28 female). The subjects represented a range of ages (from 18 to over 60), a variety of racial and ethnic backgrounds, educational levels from some high school to postgraduate work, and a range of household incomes from under $30,000 to over $60,000. The subjects were paid $40 for their participation in the 1 3/4-hour experiment.

2.2. Apparatus

A Western Electric Model 2500 telephone set (a basic touch-tone telephone with no function keys) was attached to a computer. As mentioned earlier, the 2500 set is a product of extensive human factors testing (in fact, some of the earliest formal human factors testing in the telecommunications industry), and the touch-tone pad and the 2500 handset have been used in the field extensively. In the 2500 set, the keys are approximately 1 cm by 1 cm (3/8" by 3/8") in size and are arrayed in a 4.3 cm by 5.9 cm (1 11/16" by 2 5/16") area. Each key has a slight indentation to help position the finger.

The numbers were displayed one at a time on the monitor of a personal computer. The monitor was positioned approximately 45.7 cm (18") from the subject, and the numbers appeared about 7.62 cm (3") below the eye level of the average subject. The numbers, therefore, typically appeared 30 degrees or less above the telephone set. The numbers were in 20-point type on the screen.

2.3. Design and Stimuli

Four lists of stimuli were used in the experiment; however, the lists only differed in the specific digit combinations used to create the strings used in the lists, and the randomized ordering of the strings within the lists. Each subject was only shown one list.

Each list contained 572 strings to be dialed. The goal was to create a sequence of strings that represented the diversity of alphanumeric strings customers may be dialing in the future. A pilot study was conducted to ensure an adequate number of errors would be captured for each string length and type, and to test the overall procedure, timing, instructions, and equipment. The majority of the strings consisted of random numbers that ranged from three digits to as long as 40 digits. Of these strings, the most frequently occurring were the 100 4-digit and 50 10-digit strings. These 4- and 10-digit strings were composed of random sequences of digits that were created separately for each list. Each of the digits on the keypad occurred approximately equally for the 4-digit and 10-digit strings across the lists. Other strings were counterbalanced to allow comparisons between: 1) random digit strings and strings that began with well rehearsed sub-strings (e.g., 1-800); 2) random digit strings and strings with repeated patterns of numbers (e.g., 121212); and 3) word strings of various lengths (ranging from 3-letter words to 10-letter words, and including even a few sentences of words composed of 40 letters) and the strings composed of the numbers that map to the letters. The lists were designed to support a variety of analyses that would be important for internal design decisions in the future, only a few of which will be reported here.

Table 1.   Examples of digit strings.

String Length	Example
7	248-5440
11	1-303-541-6267
4	4807
10	310-495-2822
20	406-571-8279-608-3540-634
30	241-499-2432-576-1755-706-887-311-0463
40	910-073-3264-453-1850-142-243-776-3286-877-5786-125

2.4. Procedure

At the beginning of the session, the experimenter read introductory instructions outlining the procedure to be followed.

To help the subjects become comfortable with the setting of the experiment and the equipment, the subjects began with three warm-up tasks. For one of the warm-up tasks, each subject was given a different set of five calling cards one at a time. As each card was given to the subject, the subject was asked to dial the number as if they were dialing it from home. In a second warm-up task, five names randomly drawn from a telephone book were read to the subject one at a time. As each name was read, the subject was asked to look it up in the telephone book and dial it just as they would from home. In the third warm-up task, each subject was asked to write down a maximum of five telephone numbers that they know from memory and to write down an identifier for each number. The experimenter then randomly (without replacement) selected an identifier from the list and read it. The subject was then asked to dial the number just as they might from a friend's house. This continued until the list of memorized numbers was completed.

The subjects were then given the specific instructions for the main dialing task. The subject was to pretend to be a telemarketer. They were to dial normally, but the idea was that they needed to dial as many numbers accurately as possible, since that is how a telemarketer might be paid. It is assumed that over the years people have adopted a balance of speed and accuracy that gives them an optimal level of performance.

When the handset was lifted, the screen displayed the word "Ready" for 1 second and then the number to be dialed appeared and the timer started. When the subject completed dialing the number, they were to hang up the handset, and the display was cleared. In this way, subjects stepped through the lists dialing each number string. They were also advised that if they realized that they had made a dialing mistake as they were dialing, they should just hang up and go on to the next number. This latter provision was added after the pilot to help us obtain an estimate of how often errors are self-detected during dialing when no visual feedback is provided.

3.   RESULTS

Figures 4 and 5 show the increase in error rates and number entry times as a function of string length. In both figures, the means obtained for each number length in the experiment are plotted against the predictions made from the meta-analysis (assuming linear relationships between the dependent variables and number length). An analysis of variance was performed to assess the effects of the number length and stimulus set lists on error rates. Only the effect of the length of the number was significant ($F(4, 144)=15.93$, $p<.0001$). The mean percentage errors made by subjects for the 4-, 10-, 20-, 30-, and 40-digit number strings included to assess the number length effect were 2.8 (SD=2.2), 6.25 (SD=4.8), 13.75 (SD=11.7), 16.43 (SD=15.4), and 23.50 (SD=23.0), respectively. It is worth noting that the accuracy of entering random number strings of other lengths that also were included in the lists was well-described by the linear relationship shown in Figure 4.

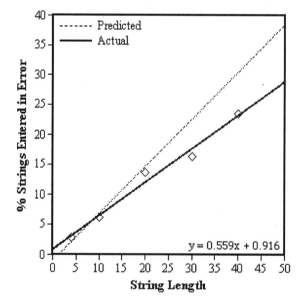

Figure 4.   Number entry accuracy as a function of digit string length.

A similar analysis was performed with entry times (the time between the first digit and the last digit entered, for strings that were correctly entered) as the dependent variable. Not too surprisingly, only the effect of number length was significant ($F(4,144)=1028.0$, $p<.0001$). The means of the entry times from the point the subject pressed the first digit of the string to the point where the last digit was dialed in a number were 1.11 (SD=.22), 4.77 (SD=.82), 11.68 (SD=2.18), 17.75 (SD=3.31), and 24.96 (SD=4.59) for the 4-, 10-, 20-, 30-, and 40-digit number strings. Interestingly, the time between seeing the number and pressing the first key also tended to increase as the length of the numbers increased (F(4,144)=14.63, p<.0001). These times were 1.28 (SD=.25), 1.21 (SD=.18), 1.39 (SD=.25), .1.30 (SD=.29), and 1.49 (SD=.39) for the shortest to longest number lengths. As a comparison, Duffy and Mercer (1978) reported a time from going off-hook to the start of dialing for 10- to 12-digit calls as averaging 1.6 seconds.

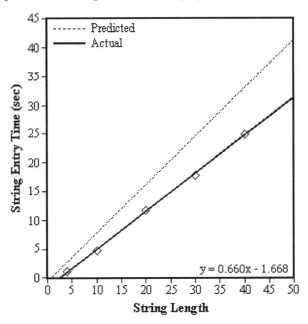

Figure 5. Number entry speed as a function of digit string length.

The average inter-digit entry times (the interval between key presses) is of more direct interest than the overall number entry times. It is the inter-digit entry time that was first hypothesized to increase as the subject entered longer strings as opposed to those strings where the digits could be kept entirely in working memory. The analysis of variance performed on the effect of string length and stimulus set list on inter-digit entry times showed this effect also was significant (F(4, 144)=339.91, p<.0001), with average times of .37 (SD=.08), .53 (SD=.09), .61 (SD=.12), .61 (SD=.11), and .64 (SD=.12) seconds as the string lengths increased from 4 digits to 40 digits. The increase in inter-digit entry times is illustrated in Figure 6. A post hoc series of Helmhert contrasts showed that the effect of number length on inter-digit entry time was due principally to the difference between the 4-digit and the longer strings, and to a lesser extent between the 10-digit and the longer strings. The remaining strings not differing significantly. Hershman and Hillix (1965) found that for the typists they tested, the average inter-digit keying interval ranged from 555 msec when three random characters were exposed at a time to 476 msec when six random characters were exposed, and from 385 msec when three letters of words were exposed down to 286 msec when six letters were exposed at a given time.

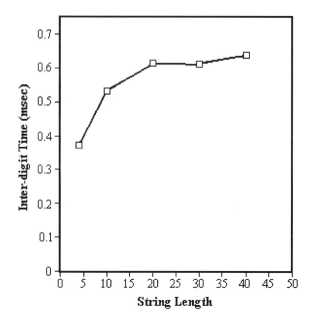

Figure 6.   Inter-digit entry time as a function of digit string length.

In the warm-up tasks, the average inter-digit dialing interval for 7-digit numbers is .52 seconds and the average interval for 11-digit numbers is .6 seconds. This corresponds reasonably well with the times found in the main part of the experiment, especially considering the poorer performance one would expect in a warm-up task.

It is reasonable to look in a little more detail at the 4-digit stimuli (100 strings were used per subject) and the 10-digit stimuli (50 strings were used). Conrad (1959) identified four main types of short-term memory errors for strings of digits, where subjects listened to strings of digits of various lengths and then demonstrated recall by entering the number on a keypad such as a touch-tone pad or through verbal recall. These include transposition errors, omission errors, substitution errors, and serial order intrusions (which in our analysis would have appeared primarily as substitution errors). Conrad found that transposition errors were the most common, accounting for 50% of errors, with omission and substitution errors being relatively less frequent (see Jurden, Laipple, and Jones (1993) for error distributions by type, gender, and age). Conrad also found that the distribution of the various types of errors tends to be in an inverted U-shape across digit positions in a string, with the most errors towards the last digits heard and the fewest errors at the beginning of each of the strings. Conrad found in one analysis that there was no apparent systematic pattern in substitutions in the noise-free system. Aaronson (1968) also found the most frequent errors to be order (presumably largely transposition) errors, and that order errors tended to exhibit an inverted U-shaped distribution across digit positions in strings. The inverted U-shape for transposition errors was attenuated for slower presentation rates and for shorter strings. Aaronson's item errors (errors of omission), however, tended not to exhibit a standard serial position effect at the slower presentation rates that would approach the continuous presence of the stimuli in our study. For the monitoring recall task, there was virtually no position effect for item errors for strings of three to four digits. While Chapanis and Moulden (1990) did not analyze their errors by class, their errors for 8-digit strings presented for five seconds do follow a monotonically increasing pattern that looks like the pattern Aaronson (1968) found (with a sudden drop in errors for the last digit position).

The distributions of the five main types of errors we found for string lengths of four and ten are shown in Figures 7 and 8. The errors are classified into transposition errors (where digits from different parts of the string, most often adjacent digits, are reversed in position), omission errors, substitution errors, insertion errors, and repetitions. Insertion errors are digits in the string that are added to the string, rather than just being substituted for another digit in a given position. Repetitions are a special kind of insertion, and so were separately classified. They appear to be an extra hit on a given key. It should be noted that the positions of some of the errors are ambiguous, as were some classifications. For example, if one digit of a pair like 22 was deleted during entry, the position of the deletion is ambiguous. For consistency, the error was assigned to the first position of the target dyad. Further, when two digits are switched, it isn't clear whether the switch should be treated as a single error or two errors. In the figure, errors were assigned to both positions held by a pair of digits that had been switched. Another source of ambiguity arises because the subjects could self-terminate when entering a string when they detected an error. If the subject entered a digit not otherwise appearing in a string and then immediately terminated entry, the error could have been one of two types. It might have been an insertion error or it might have been a substitution error. There is no way to tell with the methodology that was used. Fortunately, these kinds of errors represented a minority of the errors. For the 10-digit strings shown in Figure 8, the pattern of errors tends to look like the pattern of working memory errors for numbers reported by Chapanis and Moulden (1990). For the 4-digit numbers shown in Figure 7, the errors tend to be located on the first digit. It is not clear why this difference happened. None of the other studies have reported error patterns on strings this short before. While a high level of omission and transposition errors was expected based on the previous research on short-term memory for number strings, the high level of substitutions was surprising.

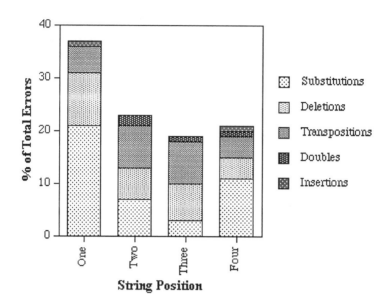

Figure 7.   Errors as a function of position in 4-digit strings.

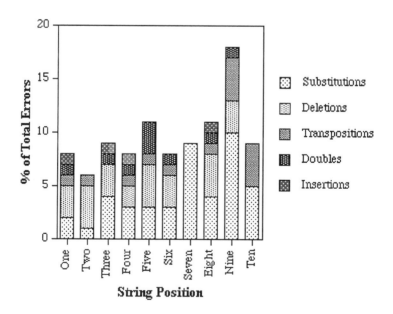

Figure 8.   Errors as a function of position in 10-digit strings.

Deininger (1960) found that over the range tested variations in keyboard layout have little effect on speed or accuracy of entry. However, Conrad's 1966 study of short-term memory effects and their relationship to the design of data-entry keyboards found that, when taken to the extreme, errors can increase 23% over those found with a key pattern that is reasonably compatible with subjects' expectations (such as the ordered sequence of keys in a touch-tone keypad). Conrad argued that the evidence did not support this difference being caused by aiming errors. Instead he argued the difference represented a decrease in recall resulting from the interference during delays incurred while the subject searches for the target and, possibly, the interference of the more difficult search itself. Thus we might expect that the differences between the 4-digit strings and 10-digit strings could be more a function of cognitive differences in how the numbers and targeting are processed than of other factors.

While a deeper analysis of the errors certainly can't be treated as definitive given the few instances in each category, there do seem to be several trends. Approximately 20% of the errors made with 10-digit strings were detected by the subjects, and the subjects then terminated entry and moved to the next string. Of the omission errors, 31% of the errors for the 10-digit strings and 17% of the errors for 4-digit strings involved deleting the number one. For 10-digit strings there seemed to be a tendency to delete one of the numbers in a doublet (27% of the deletion errors), and interestingly 32% of the 4-digit deletion errors involved deleting the first digit entered. This last error is intriguing. The equipment was well tested, and since there were both sufficient delays for it to operate effectively and the pattern was not seen in other strings, we do not believe this to be an equipment problem. It would be interesting to see if it can be found reliably in other studies. There did not seem to be any clear patterns associated with transitioning from one digit to another, beyond correctly entering the first digit of the short strings. Of the substitution errors, most involved substitutions of 7s and 4s for each other, and 9s and 6s for each other. An additional study with more stimuli in each category, resulting in more errors, would be required to understand whether the specific patterns of errors are reliable.

As Deininger (1960) noted in earlier studies, numbers with repeated characters are keyed faster and more accurately. Numbers with simple sequences of digits are keyed with similar speed and accuracy. Four digit strings such as 1111, all with the same digits, had an inter-digit timing interval on the average of 287 msec, and an error rate of only 1.25%. Ten-digit strings with alternating adjacent keys (roughly half vertical such as 2525252525 and half horizontal such as 1212121212) had an average inter-digit timing interval of 422 msec and an error rate of 6.15%. For these strings, alternating between 1 and 2 was the fastest at 396 msec. Interestingly, alternating between three adjacent digits, such as 7897897897 (again, roughly half vertical sequences and half horizontal sequences), resulted in an average inter-digit entry time of 553 msec, but a lower error rate relative to simple alternation (3.9%). It could be that the greater complexity of the pattern forced a slower entry time which, in turn, resulted in greater accuracy.

4. CONCLUSIONS

What is the source of this relatively high error rate for what would otherwise seem to be such a simple task? There is certainly a motor component to the task. It is the motor component that is demonstrated when the content to be entered is held constant and the physical characteristics of the dialing pad are varied (Deininger, 1960; Bergman, Winlow, Moore, Laidlow, and Carr, 1985). Following the meta-analysis preceding the one reported here a model based entirely on motor errors was developed that appeared to fit the data reasonably well. But if performance was entirely based on motor factors neither the effect of number length on the time between picking up the phone and entry of the first digit, nor the faster inter-digit entry times for shorter strings as compared to longer strings, would be expected.

An alternative model is that in performing the simple number dialing task a person must read some portion of the number, store it in working memory, look at a key pad (finding each number to be dialed), and dial the digits from working memory. If the full number hasn't been entered, the person needs to turn back to the stimulus, identify what has been dialed, read the next portion and store it in working memory, and so on. While some errors might be expected during the process of reading, in this model it seems likely that the biggest sources of error in a standard dialing task might be associated with the motor task of aiming for touch-tone keys and pressing them, and limitations of the working memory system. When the number is short enough to be held entirely in working memory user performance should be dominated by the motor factors. As the number begins to exceed the working memory capacity, or when something interferes with its rehearsal and maintenance, working memory errors should begin to become a factor. Further, when the subject has to shift from simply reading the number and entering it to repeatedly returning to the number to find digits not previously entered and remembering them, the time to shift, orient, and store becomes a factor in the inter-digit entry time. For small numbers this time is presumably loaded into the time before the first digit is entered.

Using a GOMS approach, a model of what might be happening during dialing can be developed. Once a chunk - a substring of three or four digits - is in working memory, the time between the digits should equal, approximately, the time to decide which digit is going to be entered next, plus the time to move the finger from whatever the current position is to the new position on the keypad and enter the digit. The former time can be approximated by Landauer's (1962) inter-digit time for silent counting of 167 msec. The latter can be estimated by calculating estimates for the times for moving between any pair of digits on the keypad, and then (for the sake of simplicity) averaging them. When two consecutive digits are the same, Card, Moran, and Newell predict that the fastest that the digits could be typed would be about 140 msec between the digits. Using the variant of Fitts's Law developed by Welford (1968), with the Middleman correction time constant (Card, Moran, and Newell estimated 100 msec/bit), predictions of movement time between any pair of keys on a touch-tone pad can be derived. Averaging these numbers yields a prediction of about 180 msec to move between keys. However, if people first visually orient to the key they are about to press with a full saccade (and if we assume that movement of the finger largely happens in parallel), the minimum movement time between digits that are not the same might better be estimated using the Card et al. estimate of 230 msec per saccade. With this assumption, using 140 msec for the same digit, a 230 msec minimum for different digits, and the Card figures for the Welford formula when the predicted movement times are greater than 230 msec, then a prediction of about 220 msec can be calculated as the average movement time between any two digits. Combining the time to select the digit with the time to "move" the finger, this suggests that smaller strings should have an inter-digit entry time of somewhere in the 350 msec to 390 msec range.

Assuming that a telephone number is placed reasonably close to the telephone (as it typically is), the time between chunks should involve a saccade from the keypad to the stimulus (230 msec), a saccade from the stimulus back to the keypad (230 msec), and the selection of a digit and movement time between keys (using 370 msec as the average of 350 and 390 msec). It should also involve finding the new chunk within the stimulus. If we assume that looking from chunk to chunk is analogous to reading and requires a series of saccades, and that each time the user looks back at the stimulus he or she searches from the beginning of the stimulus and then serially searches chunk by chunk through the stimulus, there would be an additional 230 msec for searching added to the inter-digit entry time between the first and second chunks, 460 msec between the second and third, and so on.

For a 7-digit number, then, the average inter-digit entry time would be approximately 480 msec, and for a 10-digit number (composed of three chunks), the average inter-digit entry time would be approximately 550 msec. For the 7-digit number, the extra pause between the first and second chunk of 690 msec (accounted for by the saccade to the first chunk of the number, the saccade between the first and second chunks, and the saccade back to the keypad) appears reasonably close to that reported by Deininger. Deininger's figure of inter-digit dialing times shows about a 450 msec extra pause between the third and fourth digits in a 7-digit string, although he notes that some individuals tended to refer back to the string, and some tended to simply try and remember the entire string. Those that referred back had lower errors than those who didn't, and those who didn't tended to have most of their errors towards the end of the string (in the last four digits).

However, as we move to longer strings of 20, 30, and 40 digits, the assumption of a serial search that begins with the first chunk of the stimulus and moves from chunk to chunk until it self-terminates on finding a new chunk results in predictions that move increasingly well beyond the data (with predicted average inter-digit entry times of around 670, 780, and 890 msec). A better model might be to assume that for longer strings the subject is also storing some general information about roughly where in the stimulus they last looked, and then uses two or three saccades to "zero in" on the new chunk. This would fit the gradual leveling off of the inter-digit entry times observed the longer strings, and would seem to fit the general observation of the ability people have to return to an appropriate point in a task like reading following an interruption.

If this model is correct, where might we expect errors? There should be some short-term memory errors, although these errors should tend to be lower since the chunks in general should be within working memory span. We might expect to see chunk selection errors in longer strings (e.g., a chunk being missed entirely). Also, we should see sensory-motor errors such as hitting the wrong keys. The data are consistent with the first and the last of these error types, and the rate of deletions is not inconsistent with the chunk deletion prediction. Further work to determine whether entire chunks of digits are missed, however, is needed.

As we create more complex dialing applications, therefore, we would expect always to see some level of error across a series of digits being entered. The error rate should be lower in IVR applications in which the short-term memory burden is lowered through design and by distributing the digit-entry task across a dialogue. Nevertheless, error detection and correction capabilities will always be required. Visual feedback, for example, should significantly improve the probability of entering a complete string successfully. Long strings of digits (e.g., credit card numbers) will have significantly higher error rates. Error detection and correction will be, correspondingly, even more important. Visual cues that break long numbers into "meta-chunks" (e.g., 7-digit chunks composed of 3- and 4-digit chunks) might be expected to help improve performance for these kinds of tasks.

ACKNOWLEDGEMENTS

I would like to thank Jenny DeGroot for managing the data collection at Ameritech for both the pilot and main experiments and her support during the analysis of the results. I would also like to thank Bill Cull (then a student at Loyola University) for running the pilot study, Harry Pavlidis for running the main experiment, Joe Klinger for the hardware work and some of the software work, Ray Bennett for his early software work in setting up the experiment, and the stimulating opinions from my department in general.

REFERENCES

Aaronson, D. (1968). Temporal course of perception in an immediate recall task. *Journal of Experimental Psychology*, *76*, 129-140.

Bergman, H., Winlow, T. A., Moore, T. G., Laidlow, J. B., and Carr, D. (1985). Dialing performance on Touchphone. *Proceedings of the Eleventh International Symposium on Human Factors in Telecommunications: Cesson Sevigne, France (September 9-13, 1985)*. Boston, MA: Information Gatekeepers.

Card, S. K., Moran, T. P., and Newell, A. (1983). *The psychology of human-computer interaction*. Hillsdale, NJ: Erlbaum.

Chapanis, A., and Moulden, J. V. (1990). Short-term memory for numbers. *Human Factors*, *32*(2), 123-137.

Conrad, R. (1959). Errors of immediate memory. *British Journal of Psychology*, *50*, 349-355.

Conrad, R. (1966). Short-term memory factor in the design of data-entry keyboards: An interface between short-term memory and S-R compatibility. *Journal of Applied Psychology*, *50*, 353-356.

Copping, B. (1974). Keying versus dialing: A review including an additional experiment using residential customers. In *Proceedings of the Seventh International Symposium on*

Human Factors in Telecommunications: Montreal, Canada (September 23-27, 1974) (pp. 2-10). Boston, MA: Information Gatekeepers.

Deininger, R. L. (1960). Human factors engineering studies of the design and use of pushbutton telephone sets. *The Bell System Technical Journal, 39*(4), 995-1012.

Duffy, F. P., and Mercer, R. A. (1978). A study of network performance and customer behavior during direct-distance-dialing call attempts in the U.S.A. *The Bell System Technical Journal, 57*(1), 1-33.

Hershman, R. L., and Hillix, W. A. (1965). Data processing in typing, typing rate as a function of kind of material and amount exposed. *Human Factors, 7*, 483-492.

Jurden, F. H., Laipple, J. S., and Jones, K. T. (1993). Age differences in memory-span errors: Speed or inhibitory mechanisms? *Journal of Genetic Psychology, 154*(2), 249-257.

Landauer, T. K. (1962). Rate of implicit speech. *Perception and Psychophysics, 15*, 646.

Welford, A. T. (1968). *Fundamentals of skill.* Longdon: Methuen.

Wikell, G. (1982). The layout of digits on pushbutton telephones: A review of the literature. *TELE, 34*(1), 34-45.

10 RE-ENGINEERING THE SPEECH MENU
A "Device" Approach to Interactive List-Selection
Bruce Balentine

Same Page Design Group

Key words: Menu, Point-and-Speak, Out-of-Vocabulary, List Recitation

Abstract: *List-selection is a common user activity in speech recognition applications of all kinds. Whether selecting from a short list of commands in a telephone-based IVR menu, choosing from a list of proper names in a voice-dialing application, or navigating through the n-best list returned by a large-vocabulary speech recognition event, designers often choose the menu as a well-known list-selection device. Audio menus, however, differ from visual menus in several important ways. The challenges of handling time, short-term memory, and cognitive load are issues that the speech designer must confront. This article studies the spontaneous speech acts of users confronted with verbal lists—offering specific solutions to design problems. When should the list be presented? How should user-interruption be handled? How often should the list be repeated? What are the elements of a graphical menu and how do they differ from sound? How should the menu respond to low-confidence or out-task user speech? A re-engineered menu device that addresses these audio-only challenges is presented and then discussed.*

1. INTRODUCTION

The simple act of picking from a list of options might seem to be one of the easiest actions that could be expected of a software application's human user. Indeed, the so-called "menu" in one form or another has been a commonplace device in almost every type and style of user interface for at least two decades (Rubinstein and Hersh, 1984). Users interact with menus to operate their desktop computers, check out of hotels, adjust their televisions, locate their destinations and retrieve directions to them, pay their bills, and perform a host of other daily activities with almost no attention to the underlying details of the human interface. Despite its apparent simplicity, however, the menu actually incorporates a number of subtle and interesting components that are far from simple or obvious.

Because it is so commonplace and because it is so useful, the menu deserves a re-thinking in terms of its applicability to human speech. This article provides an analysis and design for at least one interactive method—based on human speech behaviors—that takes advantage of the metaphor known as a "menu."

1.1   A Brief History of Voice Menus

So-called "voice processing" applications first began to appear in the 1980s. Although these applications used DTMF[1] as input and human speech as output, they adopted the same menu paradigm that had already been popularized on the personal computers of the day. This was partly because of contemporary biases that made the term "menu" synonymous with "user-friendly" and therefore lowered certain obstacles to product development. But it was also because of the fixed set of 12 keys on the telephone keypad—which tended to lend itself to the chunking of information that is a hallmark of the menu metaphor.

Replacing DTMF with Speech. Later, as speech recognition technologies evolved, the earliest telephone-based applications to employ speech input attempted to exploit the keyed-menu user interface model without modification. Speech was "shoe-horned" into existing voice processing applications with little or no thought as to whether the menu as a metaphor was appropriate for such a technology.

Although results were uniformly disappointing, the problem was rarely ascribed to the menu design itself. Rather, designers assumed that the speech technology would "improve" until applications became better behaved. In this context, these assumed improvements specifically meant the accurate recognition of user speech without rejection, insertion, deletion, or substitution errors. Rather than exploiting the uncertainty inherent in speech recognition as a feature of the medium, emphasis was placed on making the medium into something it cannot be. The misunderstanding continues to this day.

Confusion of Terms. As time passed, the industry came to use the term "menu" without even thinking about its implications in terms of user behavior. In fact, the term has now transformed itself to mean simply any state or node within a structured hierarchy that constrains input options at any given moment. The term "mode" is also often used as a synonym for "menu," perhaps because legal input varies from one state to the next, or perhaps because "mode" resembles "node." This confusion of terms does a disservice both to the many excellent interface designs that have nothing to do with menus, as well as to the much-maligned but often-effective menu as a simple interface device.

Objectives. This article focuses on the menu in all of its guises. From a theoretical perspective, revisiting this simple and ubiquitous metaphor is aimed at understanding how such devices can and should be constructed. From a practical viewpoint, using speech recognition to manage simple list-selection behaviors is extremely useful for a variety of applications (Balentine and Scott, 1992). This article, therefore, dissects the menu in detail with the goal of understanding deeply how the metaphor may apply to speech and how such devices may be incorporated into audio-only applications.

1.2   The "Device" Approach to Interface Design

The spoken menu discussed in this article is an extremely simple, mechanistic device. What this means is that it is deterministic in its interactions and algorithmic in conception and execution. The idea is to create a user interface that—like so many mechanical and visual devices before it—exploits spontaneous and unconscious user behaviors while remaining predictable and well bounded.

What is Meant by "Device"? Design goals for a speech interface tend to be vague and subjective, emphasizing "friendliness" or "naturalness," with little attention to explicit definitions of these concepts. For this and many other reasons, speech interface designs tend to pursue open-ended and unconstrained conversations. Such an approach has its place in certain tasks. Indeed, much progress has been made in both speech technologies and interface methods that are required for high-end applications such as travel reservations, package tracking, and stock trading. However, the assumption that such models are the ultimate goal for all speech applications has blinded designers to the value of small, deterministic, device-oriented interactions. This is unfortunate, as it prevents designers and users from taking advantage of the many benefits that derive from device design (see Table 1).

The easiest way to understand a device is to identify first what it is not. A device is not intelligent—neither in the common everyday sense, nor in the more limited sense that practitioners of artificial intelligence use the term. This means, specifically, that a device does not attempt to understand the task or the user in any fundamental way. Rather, a device presents controls to a user and then responds predictably as the user manipulates those controls. Behaviors that adjust to user error, correct user misunderstanding, or perform actions without explicit user input, when present at all, are an inherent property of design. In other words, adaptive behaviors derive directly from user input and are not perceived by the user as willful machine decisions. This distinction between mechanistic (user-dominant) and intelligent (machine-dominant) paradigms is well known in the ongoing debate over direct manipulation versus delegated interface designs (Maes and Schneiderman, 1997).

The Benefits of Device Orientation. There are in fact many good reasons for developing a deep understanding of simple, highly interactive devices such as the menu. By learning and mastering the interaction effects between a user and a machine when the protocols for turntaking, vocabulary use, and list selection are simple and formal, the student of spoken interactions develops a respect for parsimony and a focus on user goals. This rigor then serves the student well as the repertoire of techniques required for more sophisticated designs expands to address less tractable problems. Table 1 depicts the features and benefits of a "mechanistic device" which serve both user and developer. A "mechanistic device" has value in that both user and developer enjoy the benefits of simple direct manipulation. The features are, of course, interrelated. Design objectives are aimed at the most parsimonious solution to simple interaction problems.

Table 1.   Features and Benefits of Mechanistic Devices.

Feature	Benefits
Predictability	Tightly-coupled relationship between task and available user actions
	Convergence of user and machine behaviors
	Transfer of learning from one user interface component to another
Lack of Intelligence	Damping of social behaviors
	Management of the user's "theory of mind"
	Machine hard to distract; fast task completion
Generic Structure (Universality)	Reusability of similar device behaviors for similar tasks
	Fast user learning
	Clarity and purposefulness of design
Small Technology "Footprint"	Simple and inexpensive to design, develop, test and support
	High-density installations for low end-user costs
	Predictable and measurable field performance

This "mechanistic" device-orientation makes the device described here suitable for small, embedded products and for certain interactive voice response (IVR) applications that would benefit from fast and reliable list selection. Such IVR applications include voice mail, remote electronic banking, voice information services, and similar simple, well-bounded systems.

1.3   Overview of Major Issues

Phonetics and Linguistics Versus "Behaviors". The design of interactions that exploit human speech can be approached from at least three very different levels of abstraction: the acoustic/phonetic, the linguistic, and the behavioral levels.

First, all speech recognition technologies must model the *acoustic/phonetic* patterns inherent in speech. The goal, of course, is to consistently recognize these patterns whenever the user generates them. As is well known, however, such patterns are highly variant, even within the same speaker, and therefore can never be perfectly distinguished nor reliably classified without reference to some kind of extra-speech information. The classification problem is confounded when technologies must reject similar patterns that are not considered legal.

Second, many technologies incorporate *linguistic* knowledge in the form of syntax (grammars) as well as through language modeling and speech understanding methods that attempt to predict ahead of time (and to correct after-the-fact) how the acoustic/phonetic patterns of speech fit into the larger envelope of meaningful human language. Such systems are doing a very good job today of sorting through many of the problems of large vocabulary and quasi-conversational interactions. Such technologies alone, however, continue to have problems with turntaking and out-task user speech, and will soon reach a limit without reference to some sort of extra-linguistic information.

Third, the behavioral approach to interaction design—in addition to the acoustic/phonetic and linguistic results reported by various technologies—uses knowledge of basic human behaviors to infer and correct certain interaction problems. This third element of human conversation is emphasized in this article. By thinking of user input not just as a "signal to be processed" but as a behavior over which the user has only limited conscious control, a number of simple solutions can stabilize an otherwise errorful interface design. Specific user behaviors that are made tractable by this approach include turntaking, mimicking, interactive speech rhythm, and similar time-related phenomena.

'Out-Task' Vocabulary. The problem of user speech that is outside of the speech recognition vocabulary has been recognized as a problem for some years. The difficulty derives from the limitless range of expressions users may produce when confronted with a given interactive dialogue. Speech that is not represented in the recognition vocabulary is known as "out-of-vocabulary"[2] speech. Speech technologists have devoted a great deal of attention to the acoustic/phonetic distinctions that allow them to infer that such speech is under way. Many also rely on linguistic principles that allow them to anticipate, detect, and then reject (or "ignore") such user input.

Baber, Johnson, and Cleaver (1997) use the term 'Out-Task' vocabulary, a term that will be employed here to refer to this problem. The question revolves around what kinds of words or phrases spontaneously occur to users of speech applications, and whether it may be possible to anticipate the most likely as a predictive method for designing the vocabulary.

> "It would seem that choice of words for performing even quite simple tasks with speech recognition is influenced by a number of factors. The provided vocabulary plays a role but does not completely determine the choice of words. This is rather worrying in that it implies that people will attempt to use "out-task" words, or will use legal words inappropriately. The use of a speech recognition device leads to a reduction in the average length of utterances, i.e., people tend to produce shorter expressions." [Baber et al., 1997, p. 56]

Convergence. A number of investigators have observed that human speakers tend to alter their speech to make it more closely conform to the inflections, vocabulary, and rhythms of those with whom they are conversing (Baber, 1997). This tendency is known as convergence. The phenomenon is exposed in human-machine interactions when users shorten and simplify their speech when talking to a recognizer. According to Baber et al. (1997), "If people are presented with verbose feedback, they tend to use even shorter phrases than if they are presented with succinct feedback." [p. 47]

Convergence is an important concept to understand—and an important goal to establish—when designing user interfaces that take advantage of errorful and subtle technologies such as speech recognition. Indeed, its absence is more the rule than the exception. Almost all speech recognition applications appear to be quite natural and exciting as long as there are no errors. When any error appears, however, users often form a mistaken opinion of what caused the problem. They shorten their speech, experiment with alternate constructions ("I said, ..."), speak more loudly or more slowly, and answer questions with questions.

These divergent user behaviors, apparently quite spontaneous, lead to additional errors and the inevitable collapse of the fragile illusion represented by the (now stumbling) user interface. In other words, errors often lead to new errors in a cascade known as error amplification. Such divergence cannot be blamed on the speech recognition technology. Instead, the cause must be laid at the feet of the dialogue that, by allowing the user to form false impressions of both the capabilities and the internal operation of the machine, encourages users to produce highly intelligent, socially aware, and unbounded input (Reeves and Nass, 1996). The breakdown of the user interface in this context becomes inevitable because the design assumptions are inherently unstable.

The alternative, of course, is to approach the design with an eye toward *causing* convergence between user and device. That is, behaviors that are spontaneous on the part of the user should cause machine behaviors that are productive and that damp the oscillating tendency toward collapse that results from fragile design methods.

2.   WHAT IS A "MENU"?

The metaphor of the menu has become one of the most common design methods for handling command and control, navigation, fixed-data inquiry, and other "list selection" user behaviors. The metaphor is inherently selection-oriented and, of course, is originally modeled after restaurant menus (see Figure 1).

Figure 1.   The Original Menu was used as a memory aid to improve the
            human-human interface in restaurants.

As a metaphor for computer interfaces, the abstraction has survived for a quarter century. This is probably as much because it is understandable and easy to implement for designer and developer as it is because it is simple for the user. At its core, the menu is simply a printed list of options. The user reads from the list and then chooses one or more of the options. The menu is therefore an inherently visual device.

2.1   The Basic Components of a Menu

There are five basic components to a menu:

1.   Presentation of list

2.   Input pointer (with optional feedback)

3.   Selection feedback

4.   Input selector

5.   Optional feedback or corroboration interaction

The presentation of a visual menu is usually in text form. Visual menus may also be graphical in design, however, and the connection of a visual menu to its original restaurant archetype may seem so tenuous that some visual menus are not immediately recognized as such. So-called desktop metaphors, toolbars, and trays, for example, are often nothing more than menus that use graphical representations in place of, or in addition to, textual list presentation.

2.2   Visual Menus

Presentation of Options. Early menus used text-based displays for presentation and selection (see Figure 2). Each menu item would be displayed on its own line, just as with the restaurant menu. An arrow or other marker to the left—or more often an inverse video highlight—would provide the selection feedback. Later, text was replaced by graphical icons, but the basic principle of presentation and selection remained unchanged.

Figure 2.   Text-Based Menu. This menu type provided simple selection capabilities on video terminals and early personal computers. In this example of a simple DOS application launcher, the four selections are printed inside the rounded text box (*presentation of list*). The user presses the space bar—which plays the role of the *input pointer*—to sequentially highlight each text item with inverse video (*selection feedback*). The *input selector* is the Enter key, which the user presses to select the currently highlighted item. *Corroboration feedback* occurs when the user sees the launch of the desired application.

Figure 3 shows a typical pull-down menu. The design is popular in contemporary WIMP (Windows, Icons, Menus, and Pointing device) interfaces. In the figure, the presentation of options is represented by text for all eleven menu items. An additional (redundant) graphical icon is also presented for five common functions. Both are shown in Figure 3, point (A). Presumably, the icons allow the user to note those menu functions that are replicated in a toolbar (optional). As shown in Figure 4, the toolbar may be viewed as an alternate, parallel device that continuously presents the most useful subset of menu functions. Because it is not organized vertically, the eye does not immediately notice that a toolbar is, in fact, an iconic menu.

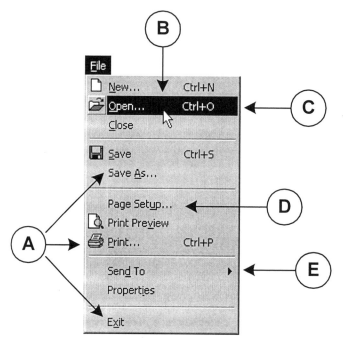

Figure 3.   Contemporary Pull-Down Menu. This menu type represents a typical example of a visual list-selection device. The *presentation of options* (A) is via text or optional icons. The *input pointer* is via pointing device with its own positional feedback (angled arrow at B). *Selection feedback* is presented through highlighting (inverse video at C). The *input selector* is a button release, click, or keypress. *Corroborative feedback* is implicit in the dismissal of the menu and the action subsequently performed by the application. Occasionally, destructive functions will use explicit corroborative dialogues—similar to the yes/no queries commonly applied to speech menus. Note the presence of submenu indicators, alerting the user to both follow-on dialogues (D) and cascaded submenus (E).

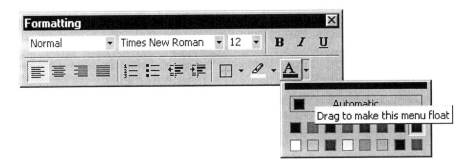

Figure 4.   Graphical Toolbar. The toolbar is not often recognized as a menu, but in fact contains all required menu elements. Symbols instead of text present available options—in this case formatting operations for word processing. There are 17 specific items shown in this example. Note the presence of submenus in the form of drop-down and tear-off menus. One—the text-color submenu—is shown opened with its "ToolTips" help exposed. The preferred input pointer and selector is a mouse or other pointing device.

Persistence and User Memory. A visual menu has the distinctive advantage of *persistence*. The display remains on the screen until the user dismisses it. The user is free to scan the menu forward and backward, considering each option without regard to the passage of time. The menu thus avoids putting any short-term memory burden on the user. This is perhaps the single most important difference between a visual and an auditory menu, one that leads to a great deal of confusion among designers.

Selection and Feedback Issues. The key to a visual menu lies in tightly coupled feedback between user action and machine action. As the user moves the mouse, for example, the pointer must instantaneously display the changing screen location. Whenever the hot zone of the mouse pointer lies within one of the menu-selection areas, the display must highlight the item. This *selection feedback,* shown in Figure 3 (C) for the *Open* function, gives the user *prior corroboration* of the task about to be accomplished.

2.3   Voice/Keypress Menus

Mapping Function and Action. When menu presentation is changed to spoken audio, the dynamic of the menu changes completely. DTMF menus use the construct "To ... Press" or "For ... Press" as the preferred list-presentation method. Although the value of this approach was not at all apparent in the early days of IVR, it is an ubiquitous design today because of its effectiveness at associating function with action in the mind of the user. Table 2 illustrates the DTMF menu.

In this *function/action* strategy, the menu presents a specific application function as clearly as possible. As shown in Table 2, for example, the machine may offer such functions as *review messages* or access to *feature options*. The presentation takes the form of an infinitive phrase—"To review your messages"—or a prepositional phrase—"For feature options"—that the user will quickly integrate and understand. This phrase is then followed immediately by the user action required to select the preceding function. The user action is an imperative active-voice construction—"Press three"—that commands the user to act.

Table 2.   DTMF Menu.

Function	User Action
"To review your messages"	"press one"
"For wakeup service options"	"press two"
"For reminder messages"	"press three"
"For feature options"	"press eight"
"To access another message box or exit voice mail"	"press star"

The DTMF menu attempts to mimic the behavior of visual menus but must compensate for a number of drawbacks. The *presentation of options* is via audio and announces each function (left column) followed by the action the user must take to select that function (right column). There is no *input pointer* and therefore can be no current selection with its associated *selection feedback*. The *input selector* is a DTMF keypress. *Corroborative feedback* is usually implied by the appearance of the ensuing function or follow-on menu. Destructive functions use explicit corroboration dialogues. Note the absence of submenu indicators, toggle states, or other enhancements common with visual menus.

The Persistence Problem. Auditory menus of course have no persistence. This, more than the often-maligned "telephone keypad," is the real reason for limiting auditory menus to a small set of possible options. In fact, users have trouble remembering more than a small number of options at a time. This problem of auditory persistence would remain if telephones had 101 keys. The challenge is one of presentation and user memory, not limitations in the input.

Auditory lists require that the user remember them.[3] This leads to an inherent conflict in terms of menu design. Should we have short and streamlined presentation methods, thereby shortening the time that the user must remember before speaking one of the items? Or should we lengthen the menu by adding explanatory detail, thereby increasing the user's understanding of the options? If we do the former, we risk the possibility that the user will commit errors by failing to understand what the options mean. However, if we do the latter we risk the possibility that the user will commit errors by forgetting what the exact action is that's expected for a given option.

Shneiderman (1997) observes that, "... voice commanding is more demanding of users' working memory than is eye-hand coordination, which is processed elsewhere in the brain." [p. 327] This is true even when a persistent (visual) list is available. So combining the short-term memory load of speaking with the additional burden of remembering the list is what makes the spoken menu defy traditional design principles. Managing time and reducing short-term memory demands are critically important if the menu, or any other interface device, is to be successfully speech-enabled.

2.4   Early Attempts At Speech Menus

Keypad Emulation. The most obvious method for applying speech to traditional IVR menus, and the first to be adopted, is shown in the left column of Table 3. In this key emulation method, the entire architecture of the menu remains unchanged. Instead, the recognizer is used to emulate the DTMF keypad, relying on the digits 0–9 plus literal words for the star and pound keys to allow the user to "speak instead of press" the telephone keys.

An alternative to keypress emulation is the use of so-called "natural words" that allow the user to refer directly to a function rather than to a key that is mapped to the function. This is superior to the use of digits, both from design and speech recognition accuracy points of view. However, the prompting method used for keypad emulation—what can be called "verbatim" prompting—takes little advantage of the new capability of such natural words. Note that a verbatim prompt is a prompt that literally tells the user what to say, e.g., "For *action* say *word*." But such prompts ignore the fact that words already have their own intrinsic meaning.

Verbatim Prompting. With "natural" (read "meaningful") words, verbatim prompting suffers from both redundancy and inefficiency. In the right column of Table 3, for example, the first menu item presents a meaningful function—*transfer to a service representative*—and then provides a user-action keyword—*transfer*—with which this function may be invoked. Although better than the use of a meaningless digit, the scheme still represents a *function/action* mapping scheme. The user must remember a structured and constrained action as the menu plays itself out, and the additional information in the prompt takes too much time for such limited utility. Designing shorter announcements becomes a major goal of this prompting method.

Table 3.   Verbatim Prompting.

Key-Emulation Method	Speech Method
"To transfer your call to a service representative, press or say *one*."	"To transfer your call to a service representative, say *transfer*."
"To hear playback of your options, press or say *three*."	"To hear your options, say *options*."
"For tips and hints, press or say *star* at any time."	"For tips and hints, say *help* or *tutorial*."
"For information about today's headlines, including international news, business, and interesting feature articles, press or say *nine*."	"For news, say *news*."

VERBATIM PROMPTING derives from a DTMF-specific presentation scheme that is ineffective with speech. The examples on the left literally prompt for "keys" that may be "pressed" by touching or speaking. Turntaking errors, illegal speech, and rejections between legal entries make this an errorful solution. The examples on the right—although seemingly benign—are long, redundant, and difficult to parse. The problem is not with the input vocabulary—which is a great improvement over digits—but with the verbatim prompting scheme. See Table 6 for an alternate solution.

To shorten verbatim prompting to its essence, the second and fourth examples use a single word for function and action. Since the vocabulary is already meaningful, however, this word is obvious and redundant, resulting in a structure that is patently silly. In the third example of Table 3, the prompt attempts to clarify the function by using synonyms for both function and action. The result is an ineffective mapping, in which redundancy is replaced by a disconnected relationship between function and action. On close examination, it becomes clear that using natural input words calls for abandoning verbatim prompting as a list announcement method.

3.   AN INTERACTIVE LIST-SELECTION ALGORITHM

Given all of the difficulties just discussed, what is the solution to the many challenges of designing a speech-only menu? The device presented here re-thinks the structure and behavior of a speech menu while attempting to retain the simplicity and basic metaphor associated with menus in general. Rather than simply copying visual or DTMF solutions, the device considers time, user memory and spontaneous speech behaviors as a starting point for effective design.

The design goals for this device include the following:

1.   Ensure that the success path provides the fastest possible task completion.

2.   Avoid presenting any unnecessary information to the user.

3.   Ensure that detours from the best-case success path return to that path as quickly as possible.

4.   Avoid dwelling on errors; as a goal, avoid declaring events as "errors" at all.

5.   Avoid expensive technology solutions, including barge-in, if possible.

6.   Use the presence or absence of user speech, even if it is not recognizable, to move the dialogue toward closure.

3.1   The Main Selection Node

The simplest speech recognition menu is shown in Figure 5. On entry to the menu, the device plays a prompt followed by an optional tone. This prompt is clear and short, allowing the knowledgeable user to speak before hearing the list of choices.

Letting the User Go First. This simplest of all speech menu designs reverses the *choice* and *list* components of the menu. The "choice"—that is, the user's opportunity to speak—is viewed as more important than the "list"—that is, the machine's presentation of options. The user who knows what to say, based on prior experience with the system, speaks immediately, achieving the highest task-completion rates (throughput). This principle is exceedingly obvious and yet is often missed by designers, who tend to focus on technology issues rather than user interactions. The philosophy can be summarized as follows: "Above all, serve the repeat user well."

As shown in Figure 5, the device plays a prompt and then activates the recognizer. The prompt must be short, and is based on the assumption that the user is expert. This basic assumption allows the user to benefit from the learning (conditioning) effects built into the interface. When the assumption is correct, the user makes a menu selection and the task is complete. The opening prompt is typically an open-ended phrase such as those shown in Table 4.

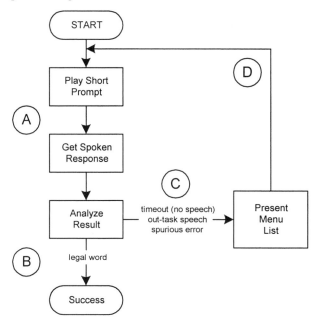

Figure 5.   User-First Strategy. This strategy presents a short prompt and then immediately allows the user to take a turn (A). By deferring the explicit listing of menu choices, the strategy rewards knowledgeable users with fast throughput. Such users follow the direct path A, B and never hear the list. Users that are unable to make a legal selection are routed automatically to the menu list (C), which takes the form of an "in-line help" routine. Because the absence of legal input automatically triggers the list, novice users perceive the behavior as "logical." Such a paradigm thus exploits all speech recognition conditions as a normal consequence of spontaneous user interaction rather than treating them as errors. After list presentation (D) the user is given another opportunity to speak (A).

The initial prompt may take several forms. The first, shown in Table 4, is the most "open-ended" and tends to be controversial among designers. The second clarifies user expectation by bounding the requested information, but at the cost of being longer. The third example takes advantage of "named menus." Since every menu has a unique name, announcing the name on entry becomes the most logical way of notifying the user and then handing over the turn. The fourth uses an upward-inflection that prompts the user for some menu element directly, in this case, one of the options for delivering a message. This last approach requires care that users can distinguish between menu prompts and yes/no questions.

Table 4.   Initial Prompt.

| "What would you like to do?" |
| "Which sports information would you like?" |
| "Main Menu" |
| "Option?" |

Detecting the Novice. If the user fails to respond to the opening prompt with legal speech, the logic does not treat this failure as an "error." Rather, it invokes a subroutine (Figure 5, point C) that presents the menu items to the user in sequential form. The decision to present the list is based on out-task rejection and similar conditions. By using such rejection as a "novice-detector" the device is attempting to exploit the "problem" of out-task speech to its advantage. Rather than try to anticipate all possible synonyms for a particular command, the design chooses a meaningful and memorable vocabulary word and then focuses on reliable rejection of all extraneous alternatives as the cue for list presentation. This logic can be summarized as follows: "Just because I don't know what the user said doesn't mean I can't move forward in a positive way." Rather than scolding the user for failing to present legal speech, the goal is to reward the user for interacting—and to introduce legal words to the user who has demonstrated a need for them—in order to encourage more informed interaction the next time around.

Note that the user may explicitly ask for the menu list, for example by saying, "What are my choices?" or, "Help, please." This input triggers the menu list, thereby empowering the user with a sense of control. The user perceives, "Whenever I don't know what to say, I can just ask for my choices. But I don't have to hear them every time the menu appears."

Conversely, the user may simply request inappropriate choices, mutter in confusion, or say nothing at all. In the face of these inputs, detected as out-task speech or timeouts, the menu engages in the same behavior, presenting the list of menu choices. The user perceives, "Whenever I'm not sure what to do next, the application helps me out. Once I know where I am, it stops talking and lets me drive."

In both cases, machine behavior is perceived by the user to be appropriate.

Detailed Discussion of the Main Selection Node. Figure 6 shows a detailed flowchart for the main selection node. The device is typically implemented as part of a library of software routines, accessed through function calls much like a high-level API.[4] Other such routines may include yes/no questions; personal name capture, enrollment, and recovery behaviors; complex numeric entry and corroboration algorithms; and similar interaction devices. At the start (Figure 6, point A) the application has specified all required control information and then invoked the menu device.

After presenting the prompt (B) the menu awaits a spoken response from the user (C). The calling routine has specified the legal vocabulary as one of its arguments. The calling routine may also set various parameters prior to calling this routine. The reader should refer to documentation specific to a given technology to understand how to specify active vocabularies state-by-state.

Continuing with Figure 6, at point (D) the recognizer has returned a legal word along with some confidence value. The user may have spoken that word or another word. It is also possible but not likely that the user has spoken an out-task expression.

The routine called *common sense* looks for a number of typical conditions to prevent runaway errors. The routine is discussed later, and has four possible outputs:

1. If return status == *good selection*, then the common sense routine has determined that the word spoken by the user has been correctly recognized by the recognizer. The menu routine returns to the application with the user's selection. In this case, note that the menu list has been skipped. This provides a user-selection path that is faster and more graceful than it would be with a DTMF-style recitation menu. The path (A-B-C-D-E) is the typical path for a repeat user who already knows the choices for a given menu.

2. If return status == *query*, then the common sense routine has determined that spoken user input is sensible but rejected. This may occur if a legal word has a slightly low confidence. It may also be that the function represented by the word is tagged as destructive, thereby requiring a query even when confidence is high. Regardless of the cause, correct machine behavior is to invoke the yes/no-question routine, thereby offering the recognized first-choice candidate to the user (F).

3. If return status == *do list*, then the common sense routine has determined that spoken user input is not sensible. Out-task speech has probably been mis-classified (insertion error). The user may also be a goat.[5] Correct machine behavior is therefore to invoke the Menu-List routine, thereby presenting menu options for the user to select (G).

4. If return status == *failure*, then the common sense routine has determined that machine and user are hopelessly unsynchronized. This condition only occurs if the user has tried once and failed, heard the list presentation, tried again, and still cannot be understood. Correct machine behavior is to fail (H), returning to the calling routine for additional help, to try alternative logic, or to switch from speech to DTMF.

Excessive background noise is usually the explanation for a failure. The condition can also occur when the recognizer keeps "hearing" a word the user has already refused via a yes/no query. Either *yes* and *no* are not being recognized properly, there is consistent misrecognition of menu words, or the user wants to be in a different part of the application and doesn't know how to get there. Regardless of the cause, this subroutine has no remaining recovery options and therefore returns without a selection. Note that this is the only way to abnormally terminate the menu subroutine.

Returning now to Figure 6, point C, the recognizer may return a response that is not a legal word (C-G). This includes a timeout with no speech detected at all, explicit recognition of out-task speech, or any of a number of spurious errors. Although we do not know what the user did, we still know what happened: the user did *not* say a legal menu selection. Rather than treat this as an error, the menu routine activates the menu-list, which replaces the role of context-sensitive help.

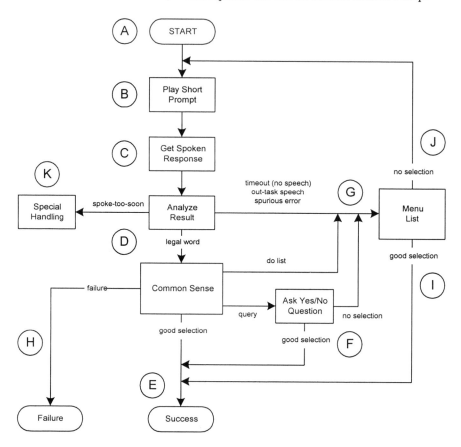

Figure 6.   Main Selection Node. The Main Selection Node is designed to be fast but stable. Expert users drop from point A to point E without even hearing the list of menu options. Novices are routed via G to the list-recitation sub-behavior. Ambiguities are detected and corrected at F. Note that the behavior is bounded—the only two exits are at E and H—resulting in a predictable device.

The Menu-List subroutine presents each menu item in sequence. This requires the same amount of time as that required to recite the list at the start of the menu, and so is not as great a penalty as help detours or other error recovery. The user may choose from menu items while list presentation is under way. There are two exits from the Menu-List routine shown in Figure 6:

1. If return status == good selection (I), then the user has interacted with the list to choose a legal option. This happens if the user speaks during Menu-List and

then optionally corroborates the selected item. As a major goal of the Menu-List routine, this results in successful termination. The function returns (E) to the application with the user's selection. This path (A-B-C-G-I-E) is the typical path for a beginning user or for a user who has forgotten the choices for a given menu.

2. If return status == no selection (J), then the user has treated the list passively rather than interacting with it. Although the user simply listens to the list without speaking, we still know that she has heard the available options. The subroutine therefore returns to the main selection node, which continues at (B) with a repetition of the prompt to speak. This must happen quickly before the user forgets the list of choices. The path (A-B-C-G-J-B-C-D-E) is thus a slightly longer success path for such passive novice users.

Referring again to Figure 6, the spoke-too-soon condition (K) represents a turntaking error. Handling the error is a special case that varies with the technology mix used by the application. Some technologies will not return a spoke-too-soon at all. Others use the condition as a feature that allows control over user turntaking (Balentine, Ayer, Miller, and Scott, 1997). Still others choose to use barge-in as a turntaking enhancement, and so need not grapple with spoke-too-soon conditions (Heins, Franzke, Durian, and Bayya, 1997). Turntaking is itself an interesting subject, but is beyond the scope of this article.

Common Sense Detailed Discussion. Even when speech recognition technology is extremely robust, certain pathological conditions can lead to repetitive failures. For example, high levels of background noise may produce endpointing errors unless and until the technology recalibrates itself. Occasional users will have one or more words that are not recognized consistently, even in the context of very small vocabularies. Such users, known as goats, represent an extremely small percentage of the population but amplify error rates whenever an application comes across them. Telephone applications are sometimes confronted with extreme dialects, foreign accents, or atypical telephone-instrument or -line conditions.

Whenever a speech recognizer experiences repetitive (as opposed to single) errors, the user invariably notes that the application has failed to exhibit any "common sense." For example, the user refuses a menu choice, is prompted to speak again, and is offered the same choice a second time. The common sense routine is aimed at detecting the obvious conditions associated with such repetitive errors, thereby damping the amplification of errors (see Figure 7).

Note that a number of special-case tests could appear in this subroutine, although the current design keeps them to a minimum. If the developer wished to create a knowledge-based—or expert system—approach to this problem, this routine is where it would go. The reader is encouraged to understand the difference between artificial intelligence and language understanding on the one hand, and this common-sense approach to minimizing error amplification on the other. The former attempts to make the machine appear intelligent to the user. The latter attempts to make the *unintelligent* machine seem *well behaved*.

Note that none of these tests are based on the acoustical properties or the meaning of the words in the vocabulary. Language understanding and AI systems would attempt such tests. This routine, instead, represents the application's *short-term memory*, keeping track of which actions have recently been completed and using that information to prevent recurrence of options that have just been refused by the user.

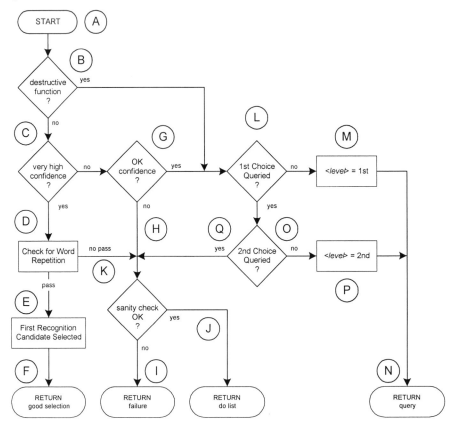

Figure 7. Common Sense logic stabilizes the menu by detecting speech and vocabulary conditions that typically produce odd machine behaviors. The routine includes a series of tests that prevent the recognizer alone from acquiring too much control over the dialogue. Repeated recognition errors, for example, are trapped here and used to avoid infinite loops. This logic represents the "short-term memory" of the menu device.

Referring to the starting point of Figure 7 (A), this routine is invoked by the Main Selection node when the recognizer returns a legal word (Figure 6, point D). The goal of the common sense routine is to determine if the input is intelligible given the immediate context of application-level events.

Referring again to Figure 7, the first test determines if this word represents a destructive operation (B). This is typically implemented with a flag that is associated with each vocabulary word. The application designer has specified this when the vocabulary itself is selected. If the operation is destructive, then the interface must ask the user to corroborate the selection regardless of any recognition confidence. Other tests are therefore not relevant. The logic will flow to point (N), leading to a yes/no query.

So-called recognition *confidence* (Figure 7, C) is a measure that varies with technology. The test at (C) simply determines how confident the recognized word is in terms of scores, probabilities, or other measures returned by the recognizer. Note that most technologies have an extremely low outright substitution rate; that is, one can expect that the wrong word will be recognized with very high confidence only rarely. Precise figures depend on the technology and the vocabulary, but would typically be under 2% of total spoken utterances. If confidence is very high, then the recognized word is almost certainly correct.

The next test therefore looks for repetitions (D). The question is, "How do we detect repeated substitutions?" This routine allows a user with a high-confidence misrecognition to fail the first time, actually receiving the menu item that was improperly recognized. However, every time this routine successfully exits at (F) a "bread crumb" is left in the form of a flag or variable setting. This allows the *word repetition* test to prevent the same error from occurring a second time. Goats, users with a specific intrusive word that is consistently recognized incorrectly, are routed to the Menu-List subroutine where they can select according to the "point and speak" method supported by the Menu-List routine (Figure 7, path D-K-G). As a result of this test, the machine will never do the same wrong thing twice in a row.

If tests (B), (C), and (D) are passed successfully, then the high-confidence recognition is allowed to stand. The first recognized word in the *n*-best recognition list is confirmed as the word spoken by the user. This straight path from (A) to (E) is thus the success path. It is also the typical path through the common sense routine. Since all tests are passed, the user has successfully selected a menu item. The common sense routine returns to the Main Selection node (Figure 6, point E), which then passes the selection to the original calling routine.

The recognized word may not have a very high confidence, but still high enough that there is good reason to suspect that the user's selection resides in either the first or second choice from the recognizer, that is, it could be an inter-word rejection. Under this condition (Figure 7, G), the best machine behavior is to ask a corroborating yes/no question, referred to here as a *query*. If, on the other hand, confidence is extremely low (H), then this is probably out-task speech that has been incorrectly classified as a word. Although a querying scheme might uncover the user's intent, it will probably be faster to send the user directly to the Menu-List subroutine.

A generic *sanity check* prevents infinite loops. If input is simply not making any sense, and this is not the first time through, the common sense routine fails (I). Otherwise, the out-task trap represented by the recognition confidence check (G-H) routes the user to the Menu-List without prejudice (J). This prevents "near-misses" from leading the dialogue astray. Details on the sanity check routine are not presented here.

Note that the *word repetition* check (D) accomplishes the same thing for high-confidence recognition. If this word has not been recognized in the recent past, then this test is passed (Figure 7, E). If, on the other hand, we have the same high-confidence recognition twice in a row, then we have a potential "repeated error" condition (K). The safest machine behavior is to invoke the Menu-List subroutine (via path D-K-J), allowing the user to hear and select from among options item-by-item.

If the recognized word represents a destructive function (B), or if confidence is not high enough to accept the word outright (C-G), then the best behavior is to ask a question. The test at (L) ensures that we have not asked this question during the current menu interaction—that is, just immediately before this recognition. If we have not (the typical case) then the common sense subroutine returns at point (N) to the Main Selection node to ask the question. The variable *<level>* (shown at point M) indicates that it is the 1st choice from the recognizer's *n*-best list that will be presented for user corroboration.

If this exact question (on the first recognized choice) has just been asked, then the recognizer may be in error. The test at (O) determines the recent-query status of the second-choice recognized word in the *n*-best list. If that word has not been queried during this instance of the menu, then the best machine behavior is to query it. The box at (P) specifies that it is the second choice in the *n*-best list that is to be queried. Return at (N) is to the Main Selection node (Figure 6), which then invokes a yes/no query at (F).

Continuing with Figure 7, at point (Q) the user has rejected both the first and the second choices from the recognizer, which means that this is the third time we've been here. Things are clearly getting nowhere. This is an extremely rare situation, which may be explained by several conditions:

1. The user may not be speaking at all, and this is actually background noise being captured consistently as speech (unlikely).

2. The user may be experiencing yes/no difficulties and has in fact accepted one of the options but lost it due to misrecognition of "yes" as "no."

3. The user is in the wrong menu and keeps saying a legal word, but not a word in the current vocabulary (most likely explanation).

4. This user is a real goat talking to a large vocabulary and getting nowhere.

Regardless of the explanation, the best machine behavior is to return to the calling routine and let higher-level algorithms figure out what to do.

Exit at point (N) means that one or more yes/no questions will be used to resolve the user's intent. The variable *<level>* points to the menu item to be queried. Note that the query tests and their alternate exits at (J) and (N) can be phrased according to this logic: "Have I asked the user if he wants this choice? If not, ask him. If so, ask him about the second choice. If both have been asked, then give up."

Every time the Main Selection node or the Menu-List subroutine asks the user a question, a flag is set that indicates which menu item was queried. These settings are too detailed to show in the subroutine flow charts. Note that all of these flags should be cleared upon exit from the menu device. Otherwise the user will not be able to select the same choice upon reentering the menu at a later point in the application.

3.2   The Interactive Menu List

As is the case with any menu device, there always comes a time when the set of legal choices must be presented formally to the user. As discussed earlier, this is more appropriately done after rather than before the user's initial opportunity to speak. Specifically, the following recognition results at the Main Selection node will trigger the menu list:

1. The explicit detection of out-task speech: a rejection with confidence so low that a query is unlikely to be fruitful (inferred out-task speech).

2. The answer "no" to queries aimed at resolving a rejected word or word pair.

3. Failure to get a reliable "yes" or "no" on a query.

4. A timeout with no speech detected.

5. Any other spurious error or condition.

   In other words, the device is not concerned with what the user said or did. Rather, it is responding to what the user did *not* do—that is, speak a clear legal expression.

The List Introduction Announcement. In the example, a "list announcement" introduces the list. The announcement represents a generic instruction—indicating that the user is expected to speak and implying that a list of options is forthcoming. The structure of the announcement therefore allows the list itself to drop the redundant "say" or "speak" verbs that are so unwieldy in the verbatim-prompt model (see Table 3). Example announcements are shown in Table 5.
   The List Introduction Announcement is a logical and forward-moving prompt regardless of the user speech that may have triggered the list. The prompt appears because the user failed to select a legal choice at the Main Selection node (Figure 6). The expressions on the left in Table 5 have been observed in demonstrations and tests but have not been quantified through a rigorous analysis. The design assumption is that users are likely to interpret the following machine speech as "logical" regardless of their own behavior, and that such an interpretation is likely to encourage convergence. The announcement's goal is to make "illegal" speech useful to the application and, apparently, productive to the user. In the table, any user phrase from the left column followed by any machine announcement from the right column to understand this design goal.

Table 5.   List Introduction Announcements.

Typical "Illegal" User Speech	Typical List Introductions
I don't know what to say.	You may say these phrases.
What are my choices?	You may say ...
Choices, please.	Please select from the following.
Help.	Please say ...
Well, what can I say?	You may choose one of these.
<no user speech at all>	Your choices are ...

Rather than attempting to bound the user's behavior by "predicting" and thereby recognizing a large set of illegal expressions, the device instead focuses its recognition task on correctly recognizing legal speech within a very constrained and small vocabulary. This makes reliable rejection of almost all illegal speech a more tractable technology challenge. It also makes such speech "quasi-legal" in the sense that it moves the dialogue forward rather than triggering "error recovery" dialogues that exasperate the user and cause divergent behaviors.

The announcement is followed immediately by a machine recitation of each legal vocabulary entry. This takes the form of a discrete menu-item playback.

Discrete Menu-Item Playback: Taking Advantage of Meaningfulness. The Menu-List algorithm presents menu items in the form of discrete spoken selections. This differs from methods commonly used for both DTMF and speech menu presentations. To understand this distinction, go back to the list-recitation methods shown in Tables 2 and 3.

Keypress-oriented prompting schemes (referred to in this article as *verbatim prompting*) work adequately when the input modality is DTMF due to the multimodal character of the interface. The presentation style is less effective for speech, however. There are several reasons why this is true:

1. The user has more trouble "parsing" the message to extract a clear and unambiguous action.

2. The explanatory speech at the front of the prompt obscures the simplicity of a spoken command. The implication is that there are far more options than are actually available.

3. The scheme does not take advantage of the mnemonic character of speech as a medium. That is, the self-explanatory "meaningfulness" that is one of the advantages of speech is not exploited in this verbatim paradigm.

In contrast, note the Menu-List design in Table 6. The device speaks each menu item in turn as a discrete word or phrase. Prompting by example is a better way to let the user know about menu options. By listening to discrete prompts that are identical to the input vocabulary, the user learns both function and action concurrently. Such a paradigm exploits the mnemonic attribute of spoken input. Note that the final prompt is not a menu element but the menu prompt at the Main Selection node.

Table 6.   Prompting by Example.

"You may chose one of these:"
"News ..."
"Sports ..."
"Weather ..."
"Which choice would you like?"

The Recognition Window: Pausing Between Items. Note that verbal recitations of any list require at least a short pause between each element in the list. This is to prevent "running together" the elements of the list and causing the user to have to think about parsing the list into individual elements. The ideal pause is probably between 500 and 750 msec. Although this is usually handled during prompt editing—that is, the digitized recording of each menu element may include considerable silence at the front and back—this menu design uses the speech recognizer to control the pacing of the list.

Each menu element is presented through a digitized recording. These recordings are carefully edited to remove all extraneous silence before and after the actual word or phrase. During playback, each menu item is followed by a brief recognition window. This menu element plus its following pause are at points (B) and (C) of Figure 8. The timeout value for the recognizer is short, typically, 1200 msec or so. In the absence of user speech, this window represents the "pacing pause" between list elements. Although it is a bit longer than the ideal, the window duration is not great enough to interfere with the user's sense of context.

Query and Selection Logic. The goal of the query and selection logic is to ask the user a simple question that is likely to resolve the user's goal as she navigates through the menu (see Figure 9).

The phrasing of the question is quite important. Many speech applications today use questions based on the top two or three choices in the n-best list. However, such questions tend to be phrased as, "Did you say ...?" Such speech-centric queries are better replaced by goal-oriented questions about what the user wants (Balentine, et al., 1997). Such questions tend to be more honest and more effective than queries about what the user has just said, as shown in Table 7. Goal-Oriented Query is a more robust method for moving the user forward. By avoiding a preoccupation with what the user said, the question makes more sense in more cases, for example, when the user "points" at a list item with out-task speech.

Table 7.   Goal-Oriented Queries.

Typical Rejection Query	Goal-Oriented Query
Did you say "News?"	Do you want news?
Did you say "Cancel?"	Do you want to return to the
Was that "Main Menu?"	Main Menu?

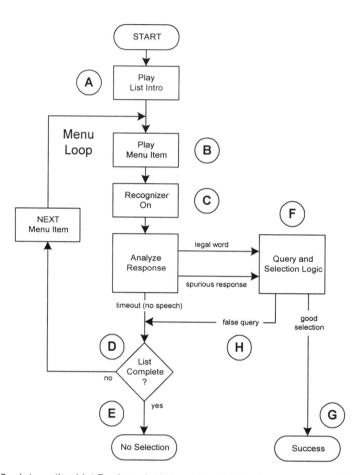

Figure 8. Interactive List Device. An interactive List Device encourages the user to speak. A list introduction such as, "You may choose one of these," appears once at the beginning of the list (A). The first menu item is then presented in isolation (B). After presentation, a short recognition window monitors the input for possible user speech (C). If the user does not speak, the recognizer typically returns a timeout. A test (D) then determines if the entire list has been presented. The sequence is repeated for each item in the menu—typically five or so choices—until the entire list is complete (E). This Menu Loop B, C, D thus offers ample opportunity for user interaction. In the absence of speech (E) the device is finished and returns to the top of the Main Selection node (Figure 6). If user speech occurs between menu items, it is detected by the recognition window and processed by the Query and Selection routine (F). The result is usually a successful selection (G) (the user has interrupted the list), and the menu is finished. False queries caused by extraneous user speech or background noise (H) return to the menu loop with minimum impact on the dialogue.

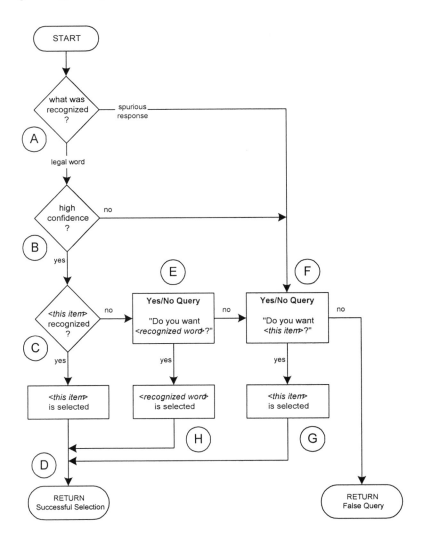

Figure 9. Query and Selection Logic. Query and selection logic resolves user goal. If a legal word is recognized (A) with high confidence (B) and it happens to be the same word as the current menu item (C), then the correspondence between speech and current list position is perfect—the user has just repeated what the machine just said. The selection is confident and no query is required (D). If input is a different legal word with high confidence (E), then the query is about the input speech—the user has probably repeated a menu item that was presented earlier in the list. All other conditions trigger a query on the current menu item—the one just spoken by the machine (F). "Yes" responses confirm that the user wants this item (G) or the one that was recognized (H). "No" responses return the user to the menu list for further interaction.

Point-and-Speak. Occasionally, users either are told explicitly or discover on their own that any speech between menu elements triggers a yes/no query on the item most recently spoken. These users choose phrases such as "that one" or "sure" to select the item just presented. This can be viewed as a "Point-and-Speak" feature that gives access to users who have trouble with recognition accuracy or who find selection behaviors convenient. An example is shown in Table 8.

Table 8.   Point and Speak.

"Point-and-Speak" Example Dialogue	Actually Recognized
User: *I don't know what to say here.*	out-of-vocabulary
App: "You may say these words." App: "News ..." App: "Sports ..."	timeout (no speech)
User: *That one ...*	out-of-vocabulary
App: Do you want sports?	
User: *Yes.*	"Yes"
App: Which sports info would you like?	
User: *Whaddya got?*	out-of-vocabulary

App: "You may choose one of these."
        "Scores ..."
        "Statistics ..."
	timeout (no speech)

User: *Bingo.*	spoke-too-soon
App: Do you want sports statistics?	Yes
User: *Yes.*	

Point and Speak Feature is an accidental but useful by-product of the list presentation design. In this example the user says "selection" words rather than repeating menu elements. Because the machine presents queries when it cannot recognize user speech, the behavior becomes one of *pointing and speaking* to choose a menu item. Although slower than the use of legal speech, the dialogue still moves users toward goals. This "goal-oriented" focus of the machine, perceived by the user as intelligent and well behaved, is an emergent property of the menu design itself and does not require the application of artificial intelligence to the problem.

4. CONCLUSIONS

4.1   Summary

Summary of Device Features. The re-engineered menu described in this article has the following specific design features:

1.   Suppressing the list until after the user has a chance to speak (user-first strategy). This benefits users in several ways:

    •   Automatic detection of experts and novices

    •   Fastest throughput for experts

- Automatic coaching for novices
2. Using out-task speech as a trigger for the list of menu items:
   - Minimum need to instruct novices
   - Logical machine behavior in the face of user uncertainty
3. Presenting menu items as a discrete list:
   - User learns vocabulary quickly
   - Fast menu presentation
4. Using the recognizer to pace menu items and to detect user-interruption:
   - Takes advantage of pauses that are required anyway
   - Allows list interruption without requiring barge-in
5. Generating queries on current menu choice when user speaks during list:
   - Allows "point and speak" user behaviors
   - Exploits spoke-too-soon and out-task speech without error recovery

These features interact to produce a device that satisfies all of the design goals described at the start of section 3.

Expert Versus Novice. The "user-first" strategy serves expert users well. By using open-ended prompts, however, and then presenting the list based on out-task speech, the device also appears to be logical to the novice. This balance between the two makes the design intrinsically stable across the entire life-cycle of use, allowing users to acquire skill at their own rate. All users are rewarded for behaviors that turn them into experts.

The real issue here is one of ease of learning versus ease of use. Jonathon Grudin (1989) notes that,

> "... ease of learning can conflict with subsequent ease of use. When this happens, priorities must be established carefully. If learning isn't possible, use will not happen. However, people buy systems and applications not to learn them, but to use them. We generally build user interfaces, not learner interfaces." [p. 1166]

The menu device described in this article is based on a design model aimed at helping a user move toward goals, a model that is automatically balanced between ease of learning and ease of use. By thinking about goals rather than interface operations, and by avoiding the help routines and user scolding that are typical of error recovery mentalities, the user is free to explore the device without pressure. Novices learn as they wish. Experts move confidently forward. The machine tends to talk more to the former, and less to the latter. But all such machine behaviors are in the framework of simply "getting to the next step." This makes for simpler coding, faster development, and more reliable testing, as well as more effective user conditioning, than with modal solutions to the novice/expert problem.

4.2   Exploiting Out-Task And Mis-Timed Speech

One of the goals of this design is to address the out-task problem directly, not by "recognizing" a large set of inputs with the goal of rejecting them, but rather using rejections themselves as positive input that accomplishes reasonable user goals without error recovery. This approach is most effective in the two major parts of the interface: the Main Selection node and the interactive Menu-List routine that presents an ordered list of menu elements.

Rejections at the Main Selection Node. The "user-first" strategy at the start of each menu allows the user to say anything essentially. Users that know the menu produce in-task expressions. By maintaining a small and highly specific vocabulary, however, the design rejects typical out-task speech with high reliability. Indeed, it is probable that small, task-specific vocabularies will always be more robust than those that include "synonyms" or explicit out-task words.

Users who know the menu selections tend to produce in-task speech as a result of convergence. Those who do not—by definition, the users that need the help—receive a list of specific menu choices. This mechanistic machine behavior tends to accelerate convergence by conditioning productive user behavior and by providing negative reinforcement for in-task, but out-of-vocabulary, speech.

One would think that such a limited and structured interaction would be rejected by users, especially given the expectations that they bring to a speech recognition interface. However, users do not report that the device is "strict" in the sense that its constraints are annoying. Rather, users tend to refer to the interaction as "intelligent" and "user-driven." This perception of user control apparently is an emergent property of the device's tendency to keep giving the turn back to the user. The machine simply presents a fact or asks a question, but the focus is always on the next action expected of the user.

Spurious Input During List Presentation. During list presentation, spoke-too-soon conditions or rejections automatically trigger queries aimed at moving the user toward a goal. In this context, phrases such as, "That one ..." or "I'd like sports" are clearly out-task expressions. It takes less time, however, to query on the current menu item than it does to "recover" such speech "errors" based on assumptions about user intent. "I'm sorry, I didn't understand" may seem to be a logical explanation. But irascible users often find such dialogues irritating and pointless. These users are then more likely to present exaggerated or angry speech to subsequent states of the application, leading to the inevitable divergence of user and recognizer. Instead, the Menu-List device makes list presentation more interactive by exploiting out-task speech to the advantage of both designer and user.

4.3   Variations And Alternate Applications

The menu device is designed primarily for simple telephone-based IVR applications. However, it has applicability in a number of other application areas.

Speaker-Dependent Personal Name Directory Listing. Voice-activated dialing (VAD) and similar applications—both telephone-based and in embedded, in-car or handheld devices—use speaker-dependent speech recognition for personal names, company names, and other user-specific data. The user speaks one of these custom entries to dial the phone, transfer calls, communicate with personal assistants, and perform similar tasks that require personalized vocabularies. The menu device described in this article may be useful for responding to lists of such names. Such "menus" of names can be quite a bit longer, as the user already knows the list.

Large Vocabulary n-Best List Selection. One of the most difficult issues associated with contemporary large-vocabulary recognition technologies is how to handle the *n*-best list. Although these technologies have evolved to the point where they exhibit remarkable accuracy on vocabularies of 5,000 and even up to 50,000 or more active words, the problem of inter-word rejection will always remain a challenge. Designers may fall back on a list-presentation device, including point and speak behaviors, to recover any word that correctly appears in the top five or so choices. If carefully designed, such interactions can be expected to be faster and more reliable than re-prompt or natural-language recovery schemes.

Toggling States. Certain applications include words that toggle the interface between two states. "On" and "Off," for example, may be used to set flags that control voice-message indicators such as *confidential* or *urgent*. Similarly, "Start" and "Stop" or other methods to activate and deactivate the speech input itself allow the user to switch between active and inactive states. Such toggling behaviors can be difficult to design, as the list presentation must explain both which options are available and what their current (toggle) state is. A listing scheme that allows the user to select an option and then toggle it is a simple variation on the menu device described here.

4.4. Drawbacks and Inapplicable Uses

The simple device described in this article is limited in a number of ways. The following specific drawbacks are areas that must await further research.

1. The design doesn't scale up well as vocabulary sizes become too large to recite. This is both because the list cannot be recited without a major time penalty, and, equally important, because rejection of out-of-vocabulary speech becomes less reliable as vocabulary size goes up.

2. The design doesn't lend itself well to data entry-intensive applications. Users who are constantly putting data into an application, for example—speaking numeric information, filling out forms, or specifying dates and times—require interactive devices that are more sophisticated than menus.

5. CONCLUSION

There are many models for speech recognition user interfaces. With the maturing of technologies today, designers have an increasingly diverse palette of design methods with which they can build usable and effective interfaces. There are many cases, however, in which modifying an existing device can provide a stable and simple dialogue that speaks more eloquently than do sophisticated and unnecessarily intelligent interactions. The menu as a list management mechanism is one such device.

Rather than dismissing simple and mechanistic devices as limiting and archaic, this paper has argued for a more vital focus on the underlying details that make such devices work well in traditional contexts. Once a designer has come to understand the operation and capabilities of a given interface design, then applying it to other media or environments can often be accomplished with only minor alterations.

By re-engineering the menu, we can breathe life into new and legacy applications.

NOTES

[1]

The acronym DTMF stands for Dual Tone Multiple Frequency—the so-called "touch-tones of the contemporary telephone keypad. This article uses DTMF as the preferred term because Touch-Tone was originally a registered trademark of AT&T.

[2] Out-of-vocabulary words (OVW) or out-of-grammar (OOG) speech is detected by the recognizer and presented to the application in different ways by various technologies. In general, it is detected at the time of recognition through a scoring process that decides—in effect—that the input is "not very close" to anything known by the machine. Some technologies explicitly report this condition; others use "low confidence" or similar measures to alert the application—which then infers that input was "illegal."

[3] It should be noted that user memory issues vary with the context of the interaction. One could argue, for example, that certain lists are already known to the user and that device behavior can be somewhat different in such cases. Lists of movies, for example, may be longer, and may be presented more rapidly, because the user already has titles in mind and is merely interested in "discarding" options that clearly are not wanted. It is not reasonable to assume that all users want to hear all lists in their entirety. Nevertheless, the basic assertion that "auditory lists require that the user remember them" is a valid one. The fact that some users bring this memory *a priori* to the interaction is a detail that impacts certain alternate designs that are beyond the scope of this article. The primary point in the context of command menus is that some users remember the options from previous encounters with the application while others do not. For those that do, any presentation is too much. For those that do not, memory issues dominate the design of list presentation.

[4]

For those readers that are not software developers, the acronym API stands for Application Program Interface, a layer of software that uncouples the application from device-specific input/output code. Speech vendors and toolkit developers are beginning to incorporate behavioral solutions such as the device described in this article into various API frameworks. By creating such API functions, these solutions decrease hardware dependencies and make for standard, reusable software.

[5]

The fanciful terms *goat* and *sheep* refer to certain voice types or speaking styles among users of speech recognition. Sheep experience few errors and, when there are errors, tend to "get it right" the next time. Conversely, goats tend to have higher substitution rates, and also experience stubbornness on the part of the recognizer—certain words simply are not recognized consistently. Although rare in the population as a whole, goats are difficult to detect. The combination of the confidence check (Common Sense) routine plus the point-and-speak feature of the interactive list attempts to serve these anomalous high-error users by getting them back on track.

REFERENCES

Baber, C. (1997). *Beyond the desktop*. San Diego, CA: Academic Press, San Diego, CA.

Baber, C., Johnson, G., and Cleaver, D. (1997). Factors affecting users' choice of words in speech-based interaction with public technology. *International Journal of Speech Technology*, 2(1), May, 45–59.

Balentine, B., Ayer, C., Miller, C., and Scott, B. (1997). Debouncing the speech button: A sliding capture window device for synchronizing turn-taking. *International Journal of Speech Technology*, 2(1), 7–19.

Balentine, B., and Scott, B. (1992, February). Goal-orientation and adaptivity in a spoken human interface. *Journal of the American Voice Input/Output Society*.

Grudin, J. (1989). The case against user interface consistency. *Communications* of the ACM, 32(10), 1164–1173.

Heins, R., Franzke, M., Durian, M., and Bayya, A. Turn-taking as a design principle for barge-in in spoken language systems. *International Journal of Speech Technology*, 2(2), 155–164.

Reeves, B., and Nass, C. (1996). *The media equation*. Cambridge, UK: CSLI Publications.

Rubinstein, R., and Hersh, H. (1984). *The human factor*. Bedford, MA: Digital Press.

Maes, P., and Shneiderman, B. (1997, November/December). Direct manipulation vs. interface agents: Excerpts from debates at IUI 97 and CHI 97. *Interactions*, 4(6), 2–61.

Shneiderman, B. (1997). *Designing the user interface* (3rd ed.). Reading, MA: Addison-Wesley.

11 IVR FOR BANKING AND RESIDENTIAL TELEPHONE SUBSCRIBERS USING STORED MESSAGES COMBINED WITH A NEW NUMBER-TO-SPEECH SYNTHESIS METHOD

Gábor Olaszy and Géza Németh*

*Phonetics Laboratory, Research Institute for Linguistics,
Hungarian Academy of Sciences*

**Dept. of Telecommunications and Telematics
Technical University of Budapest, Hungary*

Key words: Number-to-Speech, Phonetic Rules, Speech Synthesis, Message Concatenation

Abstract: *This paper describes the phonetic analysis of spoken numbers and a special approach used to achieve high quality number-to-speech (NTS) synthesis for IVR systems. The new solution provides the possibility of combining synthesized numbers with stored speech messages for professional teleinformatic applications where numbers have to be pronounced automatically (telebanking systems, ordering services, industrial information systems). Examples for English, German, Portuguese and Hungarian are given.*

1. INTRODUCTION

Numbers play a major role in many IVR systems (account numbers, time and date, billing information, prices, telephone numbers, etc.). However, their synthesis does not reach high speech quality in most cases. The problem with such IVR systems is that fixed voice messages (A) must be combined with information that changes (B). For example: *The price of the car is 23850 dollars.*, where the first part of the message represents part A and the number represents part B. In these types of messages part A appears with a good and faultless speech structure while part B has poorer speech quality because it is realised by concatenation of pre-stored number elements. A new method by which any number can be synthesized with the same high speech quality as that of any stored message has been developed and put into application in Hungary. This chapter discusses the present situation in NTS synthesis, the theory of this new approach, and the practical design. The solution uses, on the one hand, time domain waveform elements for concatenation, as traditional solutions do. On the other hand, special phonetic rules are applied to design and cut out the waveform elements of the acoustic unit inventory from spoken sample numbers. Parallel to this, a new phonetically based rule set has been developed for the concatenation of these elements during the synthesis.

The result is a very high quality number-to-speech converter where both the segmental and suprasegmental structure of the synthesized number is realised at a high quality. Therefore, callers do not perceive any difference between the pre-stored part A and the synthesized part B of the message. This chapter will also discuss quality measurements of the pronounced numbers (both the segmental and suprasegmental aspects). Furthermore, solutions regarding the NTS synthesis of Hungarian, German, English and Portuguese will be discussed.

2.   STATE OF THE ART

Teleinformatic services use mainly stored speech in many automatic information systems. This solution ensures, in most cases, a fluent pronunciation with natural prosody, i.e., high speech quality if the whole message has been recorded as one item. This means that the inherent content of the message does not change (simple, one-element message marked as A). This high speech quality cannot be kept if a so-called complex message (A+B) must be generated where prerecorded speech elements are to be concatenated, for example, by synthesising numbers (*The present balance of your account is: 23450 dollars*). In these types of complex messages part B of the message always changes. To produce this changing part, prerecorded speech parts, words and sentence parts are concatenated. In this part of the message the fluent, natural speech structure is not realised at the concatenation points. Discontinuities can be heard either in amplitude, in intonation, in speed, or in voice timbre. The result in such complex messages is a strange sounding but intelligible human voice. Users feel these discontinuities when listening to the message but cannot do anything about them because they have no influence on the design of these services.

The above-mentioned problem with speech quality comes from the fact that the speech generation part of most present IVR systems supports only simple concatenation of speech parts, but co-articulation rules and other general speech production processes (phonetic knowledge) are not taken into consideration. The reason for this situation is that software for message construction, in most cases, supports only the above-described simple method for concatenation.

Another problem with these systems is that they are designed mainly for English, and this English-based software is adapted to other languages, which often results in errors or strange pronunciation forms in the given language.

3. THE STRUCTURE OF PRESENT NTS SYSTEMS

Traditional NTS rules combine and concatenate recorded human speech elements of numbers. Such element inventories consist of 25-55 basic number elements for concatenation depending on the language (see Table 1). In the table not only the number elements, but their pronunciation is also given by phonetic transcription. The x represents any number between 1 and 9, e.g., 2x may mean 21, 22, 23, etc.

It can be seen from Table 1 that the structure and the number of the basic building elements is language-dependent. The fewest elements are needed for Hungarian (25), the highest number for Portuguese (53), out of the four languages. The traditional concatenation of these elements results in a spoken number with good segmental quality but poor prosodic quality. For example, the number 125000 is pronounced, element by element, as:

(English)	one	hundred	and	Twenty	Five	thousand
(German)	ein	hundert	fünf	und	Zwanzig	tausend
(Hungarian)	száz	huszon	öt	Ezer		
(Portuguese)	cento e	vinte	cinco	Mil		

where the pronunciation is not fluent (pauses are kept between the elements) and the natural accent and prosody structure are also missing. Pauses are needed between the elements because in most cases the building elements were recorded as stand alone units, (1,2,3,...30,40, etc.) i.e., no rhythm, intonation, or co-articulation processes of natural speech production were taken into consideration.

4. PHONETIC ANALYSIS OF THE ACOUSTIC STRUCTURE OF SPOKEN NUMBERS

Using phonetic rules in limited vocabulary synthesis systems, can improve the speech quality. A good example for this is a Dutch train timetable application (Klabbers, 1997), in which the construction of the speech units of the vocabulary was based on similar ideas to this NTS system.

If we want to produce by concatenation fluently pronounced numbers (see Figure 1) with correct accentuation, rhythm and intonation, the following features have to be taken into consideration by the last (l) and the first (f) sounds of the elements during the planning of the element inventory: 1) **continuity in pronunciation;** 2) **changing time structure for the same building elements depending on the place where the element is pronounced in the number;** 3) **co-articulation (correct spectral shape and intensity) at the boundaries of the building elements: and** 4) **accent and intonation of spoken numbers.**

Table 1.   The basic number elements in four languages for the traditional
concatenation sythesis of any cardinal number up to 999 billion.

	Basic element	English	German	Hungarian	Portuguese
1.	1	one [wʌn]	ein [ai n]	egy [eJ]	um [ũ]
2.	1--	--	eins [ai n s]	--	--
3.	1--	--	eine [ai n ə]	--	--
4.	2	two [tu:]	zwei [ts w ai]	kettõ [kɛt:ø:]	dois [dojʃ]
5.	3	three [θri:]	drei [d r ai]	három [ha:rom]	três [treʃ]
6.	4	four [fɔ:]	vier [f i: ɐ]	négy [ne:J]	quatro [kwatru]
7.	5	five [faiv]	fünf [f ɤ n f]	öt [øt]	cinco [s ĩ ku]
8.	6	six [siks]	sechs [z ɛ k s]	hat [hɔt]	seis [sɛjʃ]
9.	7	seven [sevn]	sieben [z i: b n]	hét [he:t]	sete [sɛtə]
10.	8	eight [eit]	acht [a x t]	nyolc [ɲol ts]	oito [ojtu]
11.	9	nine [nain]	neun [n ɔjn]	kilenc [kilen ts]	nove [nóvə]
12.	10	ten [ten]	zehn [ts e: n]	tíz [ti:z]	dez [dɛʃ]
13.	10x			tizen... [tizen]	
14.	11	eleven [ilevn]	elf [ɛ l f]	--	onze [õʒə]
15.	12	Twelve [twəlv]	zwölf [ts v œ l f]	--	doze [dõʒə]
16.	13	thirteen [θə:ti:n]	dreizehn	--	trese [treʒə]
17.	14	Fourteen	vierzehn [f i r ts e: n]	--	catorze [kɑtorʒə]
18.	15	Fifteen	fünfzehn	--	quinze [k ĩ ʒə]
19.	16	Sixteen	sechszehn	--	dezasseis [dəʒɑsɛjʃ]
20.	17	Seventeen	siebzehn	--	dezassete [dəʒɑsɛtə]
21.	18	Eighteen	achtzehn	--	dezoito [dəʒojtu]
22.	19	Nineteen	neunzehn	--	dezanove [dəʒɑnóvə]

	Basic element	English	German	Hungarian	Portuguese
23.	20	twenty [twenti]	zwanzig [ts van ts i ç]	húsz [hu:s]	vinte [v ĩ tə]
24.	2x			huszon.. [huson]	vinte e [v ĩ tj]
25.	30	Thirty	dreizig	harminc [hɔrmin ts]	trinta [tr ĩ tɑ]
26.	3x				trinta e [tr ĩ tɑj]
27.	40	Forty	vierzig	negyven [neʃven]	quarenta [kwɑr ẽ tɑ]
28.	4x				quarenta e
29.	50	Fifty	fünfzig	ötven	cinquenta [s ĩ kw ẽ tɑ]
30.	5x				cinquenta e
31.	60	Sixty	sechzig	hatvan [hɔtvɔn]	sessenta [səs ẽ tɑ]
32.	6x				sessenta e
33.	70	Seventy	siebzig	hetven	setenta [sət ẽ tɑ]
34.	7x				setenta e
35.	80	Eighty	achtzig	nyolcvan	oitenta [ojt ẽ tɑ]
36.	8x				oitenta e
37.	90	Ninety	neunzig	kilencven	noventa [nuv ẽ tɑ]
38.	9x				noventa e
39.	100	hundred [hʌndrəd]	hundert [hʊndɐt]	száz [sa:z]	cem [s ẽ ĩ]
40.	1xx				cento e [s ẽ tuj]
41.	200				duzentos [duʑ ẽ tuʃ]
42.	300				trezentos [treʑ ẽ tuʃ]
43.	400				quatrocentos [kwatrus ẽ tuʃ]
44.	500				quinhentos [kiɲ ẽ tuʃ]
45.	600				seiscentos [sɛjʃ ẽ tuʃ]
46.	700				setecentos [sɛtəs ẽ tuʃ]

	Basic element	English	German	Hungarian	Portuguese
47.	800				oitocentos [oitus $\widetilde{e}$ tuʃ]
48.	900				novecentos [nóvəs $\widetilde{e}$ tuʃ]
49.	1000	thousand [θauznd]	tausend [t au z n t]	ezer [ɛzɛr]	mil [mil]
50.	1000x				mil e [milj]
51.	1000000	million [miljən]	million [m i l j oː n]	millió [milioː]	milhão [milʎã $\widetilde{u}$]
52.					milhão e [milʎã $\widetilde{u}$ j]
53.					milhôes [milʎo $\widetilde{i}$ ʃ]
54.					milhôes e [milʎo $\widetilde{i}$ ʑj]
55.		billion [biljən]	milliarde [miljardə]	milliárd [miliaːrd]	bilião [biliã $\widetilde{u}$]
55.					biliôes [bili $\widetilde{i}$ ʃ]
56.	0	O [ou]	--	--	--
57.	0-	zero [ziərou]	null [nulː]	nulla [nulːɔ]	zero [zeru]
58.		and [ænd]	und [unt]	--	e [j]

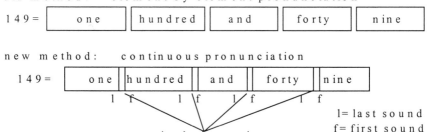

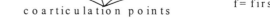

Figure 1.   Comparison of the old and the new methods of concatenation.

If these points govern the design process, the number of the units in the waveform inventory will be much higher than in traditional systems, but on the other side, a very high speech quality can be achieved. If we succeed in schematising the correct spectral shape in the joining points (last sound (l) and first one (f) of the next element), accent, timing, and intonation patterns for the synthesis, a rule system can be constructed for the concatenation, and it will be possible to realise the new method in a working form. This schematisation is language-dependent because, on the one hand, the pronunciation form of numbers may be different in other languages (see, for example, Kvale and Foldvik, 1997). On the other hand, the last and first sounds of the number elements may belong to different sound categories (vowel, consonant, voiced, unvoiced, stop, etc.). In the next sections we give a global phonetic rule set for the design, and examples will be shown as well.

4.1 Continuity in Pronunciation

Continuity and correct inherent time structure in synthesized speech (generated by concatenated elements) are important features of naturalness. The realisation of continuity physically means that the waveform elements have to be concatenated exactly by the connection points of the waveform (in general, at zero crossing points), and no pause is inserted between the concatenated elements. But this alone will not result in a continuous speech signal because correct, continuous spectral and amplitude structure is also needed at the concatenation points. Speech can be represented by a multi-variable function (spectral shape in time, inherent time structure, sound duration changes in time, average intensity of the speech wave, fundamental frequency change) where all variables have to be assigned their correct values during any kind of synthesis to achieve good quality speech. The situation is more complex if we take into account that the spectral shape itself also can be characterised as a multi-variable function in time (formants change in time). This means that continuity can be reached correctly only if the picture of spectral shape and intensity in time is realised without breaks and transients at the concatenation points of the element boundaries after the concatenation.

4.2 Changing Inherent Time Structure

By inherent time structure, we mean that the duration of the same element changes depending on its place in the pronounced number. For example, saying 131, the first 1 will not have the same duration as the second in 31. Measurements have been done for the characterisation of this phenomenon (Olaszy, 1996) by which the time structure of the building elements in more than 700 spoken numbers (containing 2 to 9 digits) was measured. The measurements showed that, minimally, 3 types of duration have to be applied for all building units if a close to natural quality pronunciation is to be reached. These are: 1) beginning (B) element (like *one* in the number 123456); 2) intermediate (M) (like *one* in the number 221456); and 3) final (L) element (like *one* in the number 7654321). This means that a correct pronunciation needs to generate these three variants in duration for all the elements which will be generated, from the point of view of co-articulation. Theoretically, every unit will have a representative for beginning (B), intermediate (M), and final (L) positions. These inherent time structure rules are not language-dependent. They can be applied for English, German, and Portuguese, too.

4.3 Co-Articulation (Correct Spectral Shape in Time)

If co-articulation is taken into consideration by system design, the effects of regressive and progressive processes of speech production have to be built into the synthesis process. In regressive processes the sound has an effect on the previous one, in progressive on the next one. During concatenation the last sound of the item and the first sound of the next item are concatenated. To realise a continuous and correct spectral shape in time at the concatenation points, the formant structure of the first and last sounds of every waveform element have to be studied and characterised. Furthermore, the other articulation factors (like nasality or assimilation between two sounds) have to be taken into consideration. When all these things are accomplished, the determination of the variants for all traditional building elements (taking into account all possible combinations) can be done.

Let us take an example for English, looking at the element *one*. This item begins with a vocoid and ends with a nasal sound. *One* can be followed (*one*XXX) by the elements *hundred, thousand, million, and billion,* (e.g., 1100) and can be preceded (XXX*one*) by the elements *thousand, million, billion, and, twenty, thirty.... ninety* (e.g., 1100, 101, or 21). To realise a **correct spectral shape** for all concatenation points of these combinations, the following variants of *one* have to be realised in the waveform unit inventory:

Rule (1) *one*, when it will be pronounced alone between pauses (e.g., 1, 2, 3), and for *one*XXX type numbers,

Rule (2) *one* (for *one hundred*), because the nasal sound will be pharyngalised by the (h),

Rule (3) *one* (for *one thousand*), because the nasal sound and the (t) have the same articulation configuration and, therefore, the (n) will be shorter than in other cases,

Rule (4) *one* (*one million*), because the (n) will change in the pronunciation into (m), for XXX*one* type numbers,

Rule (5) *one* (for *thousand one, hundred and one*, etc.), because a stop consonant is concatenated to a vocoid, and

Rule (6) *one* (for *twenty one,* etc.), because a vocoid cluster takes place by the concatenation.

So the final number of variants for *one* to realise a correct spectral shape for *one*, XXX*one*XXX, *one*XXX and XXX*one* type numbers), theoretically, will be 11 (1+5+3+2). We get these numbers from the following variants: for *one* there will be only one element, because it stays always alone, for XXX*one*XXX type numbers there are five variants as it can be seen above, for the *one*XXX type numbers there will be threevariants (rules (2), (3) and (4)), and for XXX*one* type numbers there will be two variants (rules (5) and (6)). "Theoretically" means that it is uncertain that in the practical construction of numbers every element can be used with every other element. There are cases when a certain element does not occur in XXXnumberXXX combinations. Therefore, this theoretical number may be reduced in practice (for example, *forty one* and *hundred* do not co-occur at all). When following this process for every traditional number item of Table 1, the continuity in co-articulation will be ensured at the concatenation points.

Let us take the same example for German. Rule (1) will have three variants as *ein, eins* and *eine. Ein* is pronounced in numbers greater than 1, *eins* is pronounced when it stands alone (e.g., 1, 2, 3), and eine is pronounced in the number *eine million* and in the case of *eine DM*. Rules (2), (3), and (4) will remain the same because in *ein*XXX type numbers the last and first sounds of the German number elements belong to the same categories as they do in English. The difference will be in the XXX*ein* type numbers because in German 21, 31 etc. are pronounced differently (21=*ein und zwanzig*). Therefore, rules (5) and (6) will be applied with changes as follows:

Rule (5) ein (for hundert ein, tausend ein, milliard ein), because a stop consonant is concatenated to a vowel, and

Rule (6) ein (for million ein), because a nasal sound is concatenated to a vowel.

The final conclusion for German is that, theoretically, 13 (3+5+3+2) variants of *ein* must be put into the number unit inventory for getting the correct spectral shape in every pronunciation.

For Hungarian the situation is quite different, because the number element *one* has quite different first and last sounds than in German or English. The following rules have to be applied for Hungarian *one* (*egy*):

Rule (1) egy, when it will be pronounced alone between pauses (1,2,3 etc.), and for oneXXX type numbers,

Rule (2) egy (for egy millió, like 1000000, and egy milliárd)

Rule (3) egy (for egy ezer, like 31000), because a voiced palatal sound will be concatenated with a vowel,

Rule (4) egy (for egy száz like 3125000), because the voiced palatal sound will be devocalised by the sound (s) in száz, and for XXXone numbers,

Rule (5) egy (for ....n egy, like 51, 61, 71, etc.), because the nasal sound nasalises the first sound in egy, and

Rule (6) egy (for millió egy, like 5000001), because a vowel conjunction takes place by the concatenation.

So, the final number of variants for *one* in Hungarian (to realise a correct spectral shape for *one*, *XXXoneXXX*, *oneXXX*, and *XXXone* type numbers) will be, theoretically 11, (1+5+3+2). (It will be seen in Table 4 that, practically, less than 11 variants are enough for correct processing .)

For Portuguese the design is very simple. As *one* is pronounced by one nasalised vowel, only three variants are needed:

Rule (1) um, when it will be pronounced alone between pauses (1,2,3 etc.), and for oneXXX type numbers,

Rule (2) um (for um mil like 1000000 and um bilião), because the joining (m) and
    (b) sounds have the same effect for the last sound of um, and for XXXone
    numbers, and

Rule (3) um (e.g., for vinte e um), because the transient from (j) to (u) has to be
    realised.

As can be seen from the examples, the presented method for designing the
elements is language-dependent to a certain extent. Therefore, the final number of
the variants in the waveform inventory also depends on the language. In Tables 2
and 3 we summarise the most important, language-independent, regressive and
progressive co-articulation rules which have to be used by the design.

Table 2.   The most important regressive co-articulation rules for the last and first
           sounds of the number elements.

Last sound in the Preceding element	The last sound changes to	if the first sound of the next element is
b, d, g, v, z, 3	P, t, k, f, s, ʃ	unvoiced
Ts	Ts without burst	s
T	t without burst	n
N	N(k)	k
n	N(h)	h
N	Nn	n
N	ɲ:	ɲ
N	M	m, b, p
Vowel	Transient phase	vowel
Vowel	Palatalised transient phase	palatal

Table 3.   The most important progressive co-articulation rules for the last and first
           sounds of the number elements.

if the last sound in the preceding element is	and the first sound of the next element is	the first sound of the next element changes to
Nasal	vowel	nasalised transient phase
Palatal	vowel	palatalised transient phase
Vowel	vowel	transient phase

In Table 4 we give an example algorithm (for Hungarian), showing how the main
concatenation rules are determined. To make it easier for the reader to follow these
concatenation rules, the pronounced form of each number element is given in
parentheses (e.g., 1(eJ)). The effect of co-articulation is expressed by the sound
between parentheses (e.g., (J)(ɛzɛr) means that, of the variants of the element 1000,
the variant used is the case in which the first (ɛ) sound begins with a transient part
coming from the (J) sound). The -- characters mean either the very beginning or the
very end of the number. Similar rules have been worked out for German, English
and Portuguese, as well. As can be seen from Table 4, the number of co-articulation
rules for Hungarian is about 40.

Table 4. Examples for the algorithm for the concatenation rules for Hungarian.

Previous element in concatenation	Element of the Inventory	next concatenated element	example number
--	1(eJ)	--	1
--	1(eJ)(m)	millió, milliárd	1564322
Millió	1(o)(eJ)	--	3000001
1000 (ɛzɛr)	1(eç)	100 (sa:z)	1100
..(an), ..(en), ..(on)	1(n)(eJ)	1000 (J)(ɛzɛr)	51000
..(an), ..(en), ..(on)	1(n)(eJ)	millió, milliárd	61000000
any element	2(kɛt:ø:)	100 (sa:z), millió	200, 312
any element	2(kɛt:ø:)	1000 (ø:)(ɛzɛr)	2000
any element	3(ha:rom)	100, 1000, millió	300, 3000
any element	4(ne:J)	1000 (J)(ɛzɛr)	4555
any element	4(ne:J)	millió (J)(milio:)	4000000
any element	4(ne:ç)	100 (sa:z)	400
--, 30, 100, 1000	5(øt),	100, 1000, millió	535, 5000
..(ɔn), ..(en), ..(on)	5(n)(øt)	100, 1000, millió	65, 75, 25
any element	6 (hɔt),7(he:t)	100, 1000, millió	600, 700
1000, millió	8(ɲol ts) without burst in [ts]	100	812
10--90, 100	8(ɲol ts)	1000, millió	8000, 8000000
1000, millió	9(kilɛn ts) without burst in [ts]	100	900
10-90, millió	9(kilɛn ts)	1000, millió	59000
any element	10(ti:z)	(ɛzɛr),millió,	510000
--, 100, 1000, millió	Variants for ending -ɛn in numbers1x, 4x, 5x, 7x, 9x	1, 5,	11, 115
" "	-ɛn)(k)	2, 9	12, 142, 79
" "	-ɛn)(h)	3, 6, 7	13, 53
" "	-ɛn)(n)	4	14, 94
" "	-ɛn)(ɲ)	8	18, 98
" "	20 (hu:s)	(ɛzɛr), millió,	20000
" "	Variants for ending -on in numbers 2x, (-on)	1, 5	21, 125
" "	-on)(k)	2, 9	22, 122
" "	-on)(h)	3, 6, 7	23, 1223
" "	-on)(n)	4	24, 224
" "	-on)(ɲ)	8	28
" "	30 (hɔrmin ts)	1,2,3,4,5,6,7,8,9, (ɛzɛr), millió	

previous element in concatenation	Element of the Inventory	next concatenated element	example number
" "	Variants for ending -ɔn in numbers 6x, 8x, -ɔn)	1, 5	61, 185
" "	-ɔn)(k)	2, 9	62, 289
" "	-ɔn)(h)	3, 6, 7	63, 187, 666
" "	-ɔn)(n)	4	64, 164
" "	-ɔn)(ɲ)	8	168, 968
1,2,3,4,5,6,7,8, 9, 1000	100 (sa:z)	1,4,5,8,40,50,80, 1000,millió	
" "	100 (sa:s)	2,3,6,7,9,10,1x,20 , 2x,,30,60,70,90	
5,6,7,8,9,100	1000 (ɛzɛr)	any number element	
2	(ø:)(ɛzɛr) 1000	" "	
1, 4	(J)(ɛzɛr)1000	" "	
40,50,60,70, 80,90	(n)(ɛzɛr) 1000	" "	
3	(m)(ɛzɛr)	" "	

4.4 Accent and Intonation of Spoken Numbers.

Accent and intonation also represent very important features in number pronunciation. Phonetic analysis of 300 spoken numbers showed that in the pronunciation of numbers more than one accent can be found if the number has more than two digits. For example the number 567 is pronounced as follows: **five**hundredand**six**ty**se**ven. The bolded portions indicate the places of the accents (A). The measurements also showed that accent realisation in spoken numbers can be determined by four rules. The most intensive accent (AB) is realised at the beginning of the number, a weaker accent takes place in the intermediate number elements (AM), and the weakest accent can be measured in the last number element (AL). The remaining elements in the spoken number (*hundred, thousand, million,* and *billion*) have no accent (N) at all. These four categories of accent realisation have to be realised by the concatenation synthesis as well (see Figure 2.). **Fortunately, the accent rules and the timing rules show coincidence. This means if we design elements for the realisation of the correct time structure, these elements will contain the necessary accents as well. So, either the timing rules or the accent rules have to be taken into consideration by the final design.**
   The other feature is the general melodic shape (intonation) of the number. Analysis results showed that the melody of the number represents either a falling or a level pattern. A falling pattern occurs when the number is the very last item of the message (e.g., *Your telephone bill is:* ***234 dollars***) (marked with bold letters in the example). A level pattern (bold letters in the next example) occurs when the number is in the inner part of a complex message (e.g., *The following items have been sent*

to you: 5345 and 3245 dollars). These falling and level patterns work as carrier items, and the fundamental frequency changes in accented and non-accented parts of the number are superimposed on these carrier patterns. The result is a complex intonation in the number. If a very high speech quality is to be reached in the synthesized number, both the accentuation and these carrier patterns must be realised.

Sample number	Pronounced style	comment
121	o n e h u n d r e d a n d t w e n t y o n e. AB N N AM AL	.=full stop AL= accent and falling intonation in the last item
2151	t w o t h o u s a n d o n e h u n d r e d a n d f i f t y o n e. AB N AM N N AM AL	AB, AM= accents in the number

Figure 2. The pronunciation form of the sample numbers during recording.

5. PRACTICAL DESIGN

The practical realisation of this NTS system requires the following steps: 1) determining the building elements for numbers as in the traditional systems (see Table 1); 2) designing the co-articulation rules; 3) designing the material for recording; 4) designing the algorithm for concatenation; 5) making the recording; 6) cutting out the appropriate waveform elements from the recorded numbers and placing them in the unit inventory; 7) developing the program; 8) testing the system by listening to the generated numbers and making refinements at problematic concatenation points using a waveform editor; and 9) performing listening tests. Some of these nine points have been discussed earlier; therefore, we will focus on the remaining points in the following.

5.1 Designing the Material for Recording

A set of 300 spoken numbers (maximum of 12 digits in one number) was designed, pronounced and recorded in Hungarian and in German. Experimental material was designed for English and Portuguese, too. The main goal of the design was to set up a special corpus of spoken numbers in which all the possible positions (in co-articulation, timing, accent, and intonation) occur. The design must be redundant. This means that a minimum of two examples have to be planned for every number element, to make sure that no second recording will be necessary if one element does not sound as expected. This is a very important point in the design because if a second additional recording is necessary, it is certain that the voice timbre of the announcer will not sound exactly the same as it did during the first recording. Next, in Table 5 we show examples of how the numbers containing the variants of *one* have to be designed for English. In the example, the co-articulation rules are marked with ( ) in the first column; the inherent time structure is marked by B=beginning, M=middle and L=last in the second column. The full stop (.) at the end of the sample number (in the last column) means that the numbers have to be pronounced with normal sentence intonation.

Table 5.   Example of determining the list of number elements and the source from which they will be cut (for English).

Number element	Position	rule type	Example of the recorded sample number which contains the element
one	B, L	(1)	1, 2, 1. (with pauses)
one(h)	B, M	(2)	121, 2151
one(t)	B, M	(3)	1121, 2001121
one(m)	B, M	(4)	1121151
(d)one(h)	M	(5), (2)	1122
(d)one	L	(5)	121101.
(ty)one(t)	M	(6), (3)	531231
(ty)one	L	(6)	541.

Table 5 shows that accent, intonation, and time structure are treated together in the "position" column. If we choose a B element, it will realise an accent (characteristic of the first element (AB) of the given number), the general intonation pattern, and the duration of a beginning element. The same situation occurs with the M type elements, and so on. This coincidence makes the design procedure fairly simple, i.e., only the co-articulation rules and the B, M, and L positions have to be taken into account by the design. Only the non-accented elements have to be treated as a special group in the design. These simplifications lead to 150-250 number elements in the inventory for the given language. The proper concatenation of these elements results in a very high quality pronunciation (similar to the original sample numbers that were recorded at the beginning of the work). To produce these number items, some hundred of sample numbers have to be designed (as shown in Table 5) and recorded by the selected announcer.

5.2 The Recording Session

Recording is a crucial point of the work. The announcer must be asked to utter the sample numbers with normal speed and intonation. For the realisation of the falling intonation pattern the numbers have to be pronounced as a sentence (with normal falling intonation) as described in section 4.4. For the realisation of the level pattern, sample numbers must be pronounced with non-final intonation (see the example in the first row of Table 5, in which the numbers are separated by commas). A break (1-2 seconds) must be made between the pronunciation of each number. An example shows this procedure in Figure 2.

5.3 Cutting out procedure

The appropriate waveform elements have to be cut from the recorded sample numbers and placed into the item inventory. The cutting procedure requires some phonetic knowledge. In certain cases it is advisable to follow the boundary determination by looking at the spectral picture at the given part of the number. It is advisable to cut the wave by the period boundaries at the zero crossing points. Further editing may be necessary when the system is realised and the pilot tests are performed, i.e., numbers will be automatically pronounced from the given input (typed numbers on the keyboard). A systematically performed (analysis by synthesis) test and corrections of the waveform units are advisable.

6. IMPLEMENTATION

A language-independent structure was developed for the implementation of the NTS algorithms. The block diagram of our software solution can be seen in Figure 3. It also includes illustrations for the conversion of "154."

The shaded areas are language-dependent. Blocks in the first line of the figure are implemented by program code, while blocks in the second line are data files. The input string may contain cardinal numbers (up to 999999999999) ending with a comma (,) or a full stop (.) character. The *splitter* module inserts a _ (starting sign) in front of the first digit, and generates output of a standard format (millions, thousands, hundreds, tens, and singles).

The *language-dependent processing* module omits, inserts and rearranges elements according to the requirements of the given language (e.g., in Hungarian *hundred* is used instead of *one hundred*, in German *fifty four* is expressed as *four and fifty*). The *concatenation design* function selects the required elements from the *concatenation unit inventory*. The selection is based on a three-unit-wide window (previous, current and next element) of input string (2). It is controlled by the *concatenation rules database*, which is actually a three dimensional matrix. The matrix dimensions are the elements of the input string window as is illustrated for point (2) in Figure 3. The elements of the matrix are direct references to the filenames of the concatenation unit inventory. These files contain direct speech samples. Speech output is generated by the concatenation of the unit inventory elements, prescribed in (3). This solution supports quick implementation for new languages as minimal re-coding is required. Most of the work involves waveform and ASCII file editing/modification.

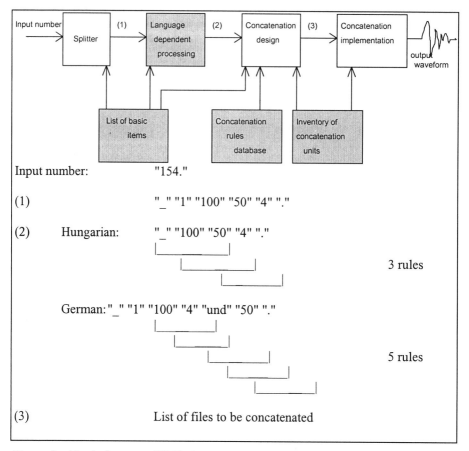

Figure 3.   Block diagram of NTS algorithm implementation.

Two systems were constructed in Hungarian for the demonstration of our method:

1) A simple number-to-speech converter PC program, which pronounces any typed-in series of numbers, separated by commas (e.g., 534, 123657, 123456789), and

2) A system for handling complex message structures (message + NTS). The complex system can generate messages like "The price of your telephone bill till today is: 13456 Forints", "The present balance of your account is: 654879 Forints", and "The last transfer from your account was: 1598765 Forints".

The non-stop demo line is available at +36-1-4631862. The female voice of the message represents a directly canned voice; the male voice in the message is produced by the new NTS system. Four topics are demonstrated: 1) bill information; 2) price of goods; 3) telephone numbers; and 4) date information. This demo system contains German, English and Portuguese number synthesis examples, as well.

7.   PERCEPTUAL TESTS

Perceptual tests were performed to evaluate the speech quality of the Hungarian system. Two tests were performed: 1) testing the speech quality of the NTS system, and 2) testing the speech quality of the message system in which a pre-stored message is followed by a synthesized number.

7.1 Quality measurements of the NTS system

Test materials have been constructed for the exact measurement of speech quality of the new NTS system. The test contained 26 numbers. Thirteen items were generated by the NTS system, and the remaining numbers were taken from sample numbers that were recorded for the NTS development. Pauses of seven seconds were inserted among the recorded numbers. Thirty test persons (adults, ages 20 to 40) were asked in two groups (n=15 for each) to listen to the material and answer two questions after listening to the number:

(1) Mark as you feel
a) the number is pronounced by a person,
b) the number is concatenated from prerecorded number elements,
c) I cannot decide?

(2) Mark the speech quality of every number on the scale --1 very bad, 2 bad, 3 acceptable, 4 good, 5 very good

The results for questions (1) and (2) are given in Table 6 and Table 7.

Table 6.   Responses for question (1).

Perceived as	Actual item: natural speech	Actual item: concatenated
Natural	302 (77%)	121 (31%)
Concatenated	--	132 (34%)
Undecided	88 (23%)	137 (35%)
Total	390	390

The results for question (1) show that, of the responses for the concatenated numbers, 31% were marked as a number pronounced by a person, and 35% were "no decision". This 35% shows that test persons who marked this answer could not hear the difference between the synthesized and human-pronounced items. On the other hand, among the responses for the natural pronounced items, 23% were given as "no decision". This means that the listeners also could not hear the difference between the synthesized and human-pronounced items. The summarized result is that, of the total of 780 responses, 346 (88+121+137) reflected no perceived difference between the natural and the concatenated items. If we take into consideration that listeners listened 390 times to concatenated items during the test, and in 31% of the responses they judged the item as natural, and in 35% the response was "no decision", the final conclusion is that 66 % of the responses signified that the speech sounded natural.

Table 7.   Judgments of speech quality.

	responses (%)
1 very bad,	-
2 bad,	-
3 acceptable,	4 (0.5)
4 good,	82 (10.5)
5 very good	694 (89)

The results for question (2) show that 89% of the responses were "very good", and 10.5% were "good" speech quality. This shows that the speech quality of the new NTS system is very good.

7.2   Quality Measurements of Messages Containing Recorded Natural Speech Combined With Synthesized Numbers

In this test the speech quality of complex messages characteristic of IVR systems was measured. Twelve complex messages were created in which a female voice (recorded) presented the first part of the message, and the second part contained the synthesized number spoken with a male voice. Thirty other test persons (adults, ages 20 to 40) were asked to listen to the items and answer the question: mark the speech quality of this complex message on the scale --1 very bad, 2 bad, 3 acceptable, 4 good, 5 very good. The test results are given in Table 8.

Table 8.   Speech quality for complex messages.

quality	responses (%)
1 very bad,	--
2 bad,	--
3 acceptable,	--
4 good,	14 (4)
5 very good	346 (96)

This result shows that 96% of the responses noted the speech quality of the complex message as very good so, practically, the subjects did not hear relevant differences between the recorded speech and the synthesized part. It means that the new NTS system produces high quality number pronunciation and, therefore, can be used successfully for complex messages in IVR systems.

8. CONCLUSIONS

This phonetically governed approach can improve the speech quality of synthesized numbers produced by concatenation of number elements. The solution requires phonetic analysis of the number elements of the given language. The result of the analysis leads to: 1) a higher number (about 250) of waveform elements in the inventory for concatenation, and 2) a phonetically based rule system for the concatenation of the elements. The realised systems show that the NTS system produces practically the same high quality speech as a pre-recorded message-based system, so it can be used to build combined messages for IVR systems.

REFERENCES

Klabbers, E. A. M. (1997, September). High quality speech output generation through advanced phrase concatenation. *Proceedings of Speech Technology in the Public Telephone Network Workshop, COST Telecom Actions 249, 250 and 258* (pp. 85-88), September 26–27, Rhodes, Greece.

Kvale K., and Foldvik, A. K. (1997, September). Four-and-twenty, twenty-four, What's in a number? *Proceedings of Eurospeech '97, Vol. 2*, (pp. 729-732), September 22-25, Rhodes, Greece.

Olaszy, G. (1996). Phonetic analysis of number elements. In Mária Gósy (Ed.), *Speech research 96* (pp. 97-109). Budapest: MTA. (in Hungarian: Számelemek kiejtésének fonetikai vizsgálata, Beszédkutatás 96 Szerk. Gósy Mária, MTA Nyelvtudományi Intézet).

12 VOICE MESSAGING USER INTERFACE

Harry E. Blanchard and Steven H. Lewis

AT&T Labs

Key words: Voice Mail, Voice Messaging, Unified Messaging, Universal Mailbox

Abstract: Voice mail is certainly one of the most common interactive voice response
(IVR) applications in telephony. When we talk of designing these systems to
be easy to use, there is much overlap with general IVR system design, but
there are also significant special characteristics to voice mail systems. This
chapter discusses designing voice mail systems for usability, with particular
emphasis on the special characteristics, demands, conventions, and standards
associated with voice mail. In addition to coverage of the ubiquitous touch-
tone user interfaces, this chapter also discusses the state of the art in voice
mail user interfaces: novel prompt and menu structures, the use of automatic
speech recognition, unified messaging ("universal mailbox"), and multimedia
mail.

1. INTRODUCTION

Voice messaging, or voice mail, is a computer application integrated with the voice
telephone network that allows people to receive, send, and store recorded audio
messages. From modest beginnings in the early 1980s as a replacement for
answering machines in business environments, voice messaging has grown to
become an essential business tool and an increasingly common consumer service.
At the end of 1998, a good estimate of the popularity of voice messaging can be
seen in the U.S. number of mailboxes, with about 60 million business "seats" with
mailboxes, an additional 12 million or so residence lines with call answering
mailboxes, and some six million or so mailboxes behind cell phones and pagers.
Numbers for the rest of the world are harder to come by, but the total international
market for voice messaging has been estimated by some of the equipment makers as
about as large as the U.S. market. Millions of calls each day are answered by voice
messaging applications and retrieved by mailbox subscribers.

The typical voice messaging application is controlled by users entirely from a conventional touch-tone telephone. Thus, voice-messaging applications fall into the class of interactive voice response (IVR) services, most of which have touch-tone user interfaces (see Chapter 7). As many have noted, the touch-tone, or DTMF, user interface is impoverished, both in the user input, which is limited to the 12 keys on the standard touch-tone telephone, and in the system output, which places large processing demands on users because of its serial audio nature.

It is useful to think of voice messaging as services built from a set of messaging capabilities. These include

1. *Call answering:* When the user doesn't answer a phone call or is busy on another call, the voice mail system answers the call and can take a message. This function is parallel to the capability provided by a telephone answering machine, except that the voice mail system can take messages while the user is talking on the telephone and is able to provide the caller with various options.

2. *Message retrieval and management:* As with answering machines, voice mail allows users to play messages left for them, and it allows them to store those messages and play them back later.

3. *Sending messages:* Voice mail systems can provide a functional analogue to e-mail that an answering machine cannot: the ability to record messages in the voice mail application and send them directly to other user's mailboxes on the same or connected systems. Sending can be invoked by subscribers who originate messages or who reply to or forward messages they receive.

4. *Call delivery:* Another form of voice messaging intercepts calls in the network that are not answered, or the line is busy, and offers the caller a chance to record a message to be delivered later. The system then keeps trying the line and delivers the message to a person or an answering machine when the line becomes free.

5. *Bulletin boards*: In this form of voice messaging, mailboxes do not belong to individual subscribers but rather are used by many people to leave and retrieve semi-public messages. This form of voice messaging is not common in the U.S.

Most business voice messaging applications provide the first three of these capabilities to subscribers using the public telephone network or private (PBX) telephone systems. Residential mailboxes typically enable call answering and message retrieval; the capability for sending messages is usually not available until there is a sufficient subscriber base. Call delivery is a form of voice messaging that doesn't assume a mailbox. Increasingly, in new services, combinations of these functions have been embedded as the messaging component of even more feature-rich systems that provide call management and intelligent agent capabilities.

Voice mail systems provide a special challenge to the user interface designer beyond the challenges of designing other IVR systems. Voice mail is distinct from call-direction and informational IVR systems in several respects:

♦ Voice mail systems are highly complex and feature-rich, and it is difficult to reduce the complexity of the system by not providing these features. At any one point or menu in the user-system dialogue, the number of choices the user has available may outstrip the dozen keys available for the user to make a response. This is particularly true in dealing with a message, both while listening to it (when the "menu" of choices is implicit) and immediately afterwards. The common strategy of breaking up functions into a series of hierarchical menus cannot always be done, e.g., when responding to a message, many functions must be available immediately after listening to the message.

♦ Unlike most IVR systems, voice mail systems are used day in and day out, often several times a day. Many users are highly practiced and need to accomplish transactions quickly and efficiently. Nevertheless, there are also novice users, who are faced with a necessarily complex, feature-rich user interface to master.

♦ In addition to the usual prompted menus common to IVR applications, voice-messaging systems need to allow users to navigate around the system freely and operate on voice and DTMF data. This often leads to more use of common commands, often invoked with multiple, sequential keypresses.

♦ A voice mailbox has two distinct sets of users. There is the mailbox box owner, often called the *subscriber*, who gets and manages messages every day. Then, there is the *caller* who leaves a message for the mailbox owner using call answering. Many callers are *not* subscribers to the particular voice mail system with which they must interact to leave a message. Callers come into the call answering user interface as the ultimate novices, with a set of expectations based upon the conventions of making a phone call and using answering machines and a variety of voice mail systems from different vendors which they encounter. Attention to these user expectations and the encouragement of cross-application standards are particularly important.

♦ Standards and consistency of design are highly significant to voice mail use. Besides the needs of callers leaving a message, described above, subscribers are also facing needs for consistency. There are an increasing number of users of voice mail who have one mailbox for their work phone and another mailbox for the home telephone. And, because of the development of the telephone industry, the office and home voice mail systems are almost always from different manufacturers, with different user interfaces and different sets of functions.

Over the last few years, we have been involved in the design and competitive analysis of residential, business, and cellular voice messaging user interfaces. In this chapter we will discuss voice mail user interface design with a focus on design decisions that are significant to voice mail systems. Readers should also be familiar with the general guidelines for IVR and touch-tone system design, as discussed by Gardner-Bonneau in this book (see Chapter 7).

2.   THE TOUCH-TONE VOICE MAIL USER INTERFACE

The touch-tone user interface for creating and retrieving messages will be the most familiar to the business user and the dominant interface for messaging for the foreseeable future. The touch-tone user interface is series of *transactions* in which the system provides audio instructions to the user, who directs the system by pressing a key or sequence of keys on the touch-tone, or DTMF, telephone keypad, or by recording speech.

The dominant metaphor for voice mail has evolved. Early voice mail systems used an answering machine metaphor in the design and presentation of the user interface. At the time, most answering machines used magnetic audiotape for recording and reflected a tape recorder metaphor. Early voice mail systems used tape recorder terminology, offering options such as "rewinding" or "erasing" voice messages. Current voice mail systems downplay the tape answering machine metaphor, even as answering machines themselves are increasingly likely to use digital, tapeless recording technology. Today, the parallels with e-mail are more compelling, so terms such as "delete" and "forward," familiar to e-mail users, are becoming more common in voice mail. The e-mail metaphor for voice mail is likely to persist, considering the burgeoning use of Internet e-mail by a wide and diverse population of users and the convergence of voice mail and e-mail into unified messaging and universal mailboxes.

2.1   Common Elements of Touch-tone Transactions

Touch-tone user interfaces can be usefully analyzed using the general user interface categories of system output, user input, and user-system dialog. For touch-tone user interfaces, system output is made up of announcements, including menus and prompts; user input is single or sequential DTMF keypresses, voice recordings, or no action; and dialog includes interactions between input and output, including timeouts and error handling. From a task analysis perspective, the caller's focus is leaving messages, which can be done by simply speaking and then hanging up, but can also include some manipulation of the voice input. The subscriber's focus is on retrieving and managing messages, as well as sending them.

Several good summaries of the general principles for designing touch-tone IVR dialogs are available (Halstead-Nusslach, 1989; ISO/IEC 13714, 1995; Marics and Englebeck, 1998; Schumacher, Hardzinski, & Schwartz, 1995). Although voice messaging differs from generic touch-tone IVR applications in many important ways, our experience has been that generic touch-tone design guidelines work.

2.1.1   Prompts

Voice messaging application trees can be "deep" and "broad," but at most points in the dialog people usually have a good idea of what they want to do. With many menu options, the convention of first stating the action, then the required key to perform that action (e.g., To do X, press Y) applies. For example:

To get your messages, press 1,
To send a message, press 2,
To record or change your greeting, press 3.

The action-response order of each of the three prompts in this menu reduces the mental work needed. The action-response order is consistent with what Clark and Haviland (1977) called the "given-new contract" in conversation and leads to faster responses and fewer errors (Englebeck and Roberts, 1990).

However, there are cases where the response-action phrasing sounds better; exceptions to the general rule may apply when there are only one or two options, for example:

Press 1 for yes or 2 for no please.

One prompt strategy that balances the needs of experienced users with those of novices is to use staged prompts, where an initial terse prompt is followed by a short embedded pause and then a more extended prompt (Brems, Rabin, and Waggett, 1995; Leopold and van Nes, 1984). Thus, for example, the prompt for addressing a message might be:

To what number? ...
[2 seconds of silence]...
Enter a mailbox number followed by the pound key.

2.1.2   Interruptibility

New users are likely to wait until all the options are played before selecting a key to choose an option in a menu. However, voice messaging applications wouldn't be tolerated by experienced users unless they allowed menu-selection keypresses before the end of the entire menu prompt, and also allowed users to provide keypresses that apply several menus or system states ahead of the playing prompt. Indeed many experienced voice mail subscribers get messages by giving their identification (ID), their password, whatever delimiters are needed, and the main menu key that starts messages playing as if it were one long DTMF string

The capability to press a key before the menu has finished playing and have the system immediately stop playing the prompt or announcement and act on this input is called *dial-through*. Unlike an IVR system that might be used once or occasionally, dial-through is absolutely essential to voice mail users. For example:

System: To get your
User interrupts: 1
System: First message ...

Dial-ahead, which is the entry of a sequence of commands before the prompts have been played by the system, is also essential. When voice mail designers choose to disable dial-through and dial-ahead, for example, to let users know if they have system or priority messages or are running low on storage, users are likely to complain.

2.1.3   Time-outs and Re-prompts

When no response is detected from the user after some time-out period, the system should respond. Systems have two major time-outs: the *time-out for no action* and the *inter-digit time-out.* The time-out for no action is the maximum time the system waits for any response after a prompt. If this time-out is too long, the user often doesn't know what is happening; if it is too short, the user can feel rushed. In our experience, a time-out for no action of 5-6 seconds works well. The inter-digit time-out is the maximum time the system should wait between keypresses when more than one digit is expected, for example, when the user is entering a telephone number. In regular telephone dialing, this time-out is 15 seconds. In voice messaging and other IVR applications, our experience has been that a time-out of 5-6 seconds works here as well.

A time-out for no action almost always should result in a re-prompt for user action. The re-prompt should rephrase the focal options and may present all of the options available, including hidden or fallback options (such as returning to the main menu). A better re-prompt might orient users by telling them where they are in the menu hierarchy or what task they are doing. For example:

> [User times out to menu following message.]
> System: You are listening to a new message.
> To replay the message, press 1 ...

After repeated time-outs of this sort (typically, three is a good number) the system should do something other than re-prompt. First, it should inform the user that a special action is going to occur, and then perform that action, such as disconnecting or connecting to an attendant.

What the system should do when an interdigit time-out expires depends upon the context in which it happens and the system capabilities. It is almost always useful to interpret the time-out as a delimiter. The system should attempt to process the data, accept it if it can be interpreted as valid, or prompt for more information, or for re-entry, depending upon the context and sophistication of the software.

So, for example, the user may enter seven digits when a telephone number is requested and time-out. The system may then interpret the number by assuming the three-digit area code, or prompt the user with:

> You must enter a 10-digit number beginning with the area code.

Two other time-outs are a time-out when the user is providing speech input and the system has detected a period of silence, and a time-out after the system has prompted for touch-tone string input, such as an address or a user ID, and no initial digit has been provided. While these are conceptually distinct types, in our experience the 5-6 second time-out works in these cases as well.

2.1.4   Feedback

Any user action or response must result in feedback from the system. Feedback to valid input is usually the progression of the system to the next prompt, menu, or message. Feedback does not mean that the requested operation is completed. If it is not, the system should inform users, after three seconds of no response from the system, that their request is still being processed. The cases that challenge timely feedback in systems usually involve extensive database access, which in voice mail systems may be address lookups.

2.1.5   Feedback to Errors

When a user provides incorrect input, as defined by the system, the system should inform the user of the error in a polite fashion that does not blame the user. Words such as "*Invalid entry*" are abrupt and not very informative. As with re-prompts to time-outs, the system should orient users by reminding them of what menu they are listening to or what task they are doing, followed by a full list of options that are available. For example:

> Five doesn't work here. You are in the main menu. To get your messages, press 1. To send a message, press 2. ...

2.1.6   Menu Length

A generally accepted principle of touch-tone design is to limit the number of options in a menu to approximately four (Englebeck and Roberts, 1990). Due to the feature richness of voice mail, fitting four options to a menu can be a challenge or an impossibility. For example, in the menu that is offered to the user after he or she listens to a message, the user has to be given all the options for dealing with the message. This includes deleting the message, replying to the message, forwarding the message, saving and/or skipping the message, and replaying the message. Often, other options will also apply. Splitting these functions at this point can itself be a serious usability problem, so, despite the complexity of the menu, the system should offer all the message disposition options. In the end, the classical consideration for trade-off between depth and breath of menus applies (e.g., Lee and MacGregor, 1985; Norman, 1991).

2.1.7   Mapping of Keys to Options

Voice mail menus should follow the principles used for design of any touch-tone menu system. Key mappings are driven by a small set of guidelines that make good sense individually but can lead to conflicts when taken as a set.

Generally, the most frequently used or default action should be assigned to the 1 key, the next most frequent to 2, and so on (Marics and Englebeck, 1998).   An exception to this rule concerns conventions or standards for key assignments. In particular, if a prompt requests a yes/no answer, the 1 key should be "yes" and the 2 key should be "no."

Menu options should be numbered, if possible, without gaps in the ordering. That is, options in a menu should be numbered consecutively from 1 to 9; the first option is 1, the second option is 2, etc. This aligns with user expectations. If a gap appears, users may overlook the renumbering, or think they missed an option while they were listening. If options are presented out of order, then users may lose track of the key-action mappings.

This rule always involves a trade-off with other factors. For example, a message disposition menu appears after the user listens to a message. Say, for example, option 5 in the menu forwards a message. If the user receives a "private" message (A "private" message, in the parlance of most voice mail feature descriptions, is a message which cannot be forwarded to a third person.), option 5 should not be offered. Rather than moving option 6 to option 5 in the message disposition menu for a private message, it makes more sense simply to skip option 5 for these messages (i.e., to offer option 4, then option 6), in order to keep the canonical assignments the user learns for this one menu. It is probably particularly important that the mappings of keys for "destructive" or irreversible actions, such as erasing things or canceling actions, be consistent across menus.

2.1.8   Global Commands

Global commands are keypresses that are common to any menu in the user dialogue, and may be used to invoke the same or similar features. One advantage of defining global commands is that they provide users with a tactic to universally "escape" out of situations in which they are confused, e.g., by going to a main menu or getting help. ISO/IEC 13714 specifies a "control menu" which is a set of options that are active at all points in the voice mail user interface. Control menu options, as defined by the standard, include:

♦   Going to the top level menu

♦   Selecting a language

♦   Forcing the system to disconnect

♦   Canceling an action or backing up to a previous menu

♦   Invoking an on-line help announcement

♦   Returning to a prior context

The advantage of specifying global commands or a control menu is that users can learn a specific key sequence for a common activity that need not be prompted.

2.1.9   Use of the "#" and "*" Keys

The "#" key is perhaps the key with the most fixed conventional meaning of any of the touch-tone keys, and has widely accepted conventions of usage. The "#" key is used to indicate delimiting or completion, and moving ahead in the system. The "#" key should be used to indicate the end of a variable-length string of numbers, e.g., the end of a telephone number or credit card number. The "#" key should be used to indicate the end of voice input, i.e., to indicate the end of a recording. It is also used to skip to the next message in a queue and by callers to skip the subscriber's outgoing call answering greeting and go directly to the record tone. If "#" is used in other circumstances, it is recommended that it be used to indicate completion of an on-going action or entry.

Two common conventions appear for the "*" key. One convention reserves the key entirely for canceling user input or task steps or for moving back to the previous logical option or choice in the call flow. The other convention uses the key as a way to access a further set of menu options beyond the one to nine single-digit options.

Voice mail systems often have a large number of options, and voice mail often makes use of global options that pertain to all menus, in other words, a "control menu." Because of these features, it makes more sense to choose the convention which allows the "*" key to access more options, rather than restrict it to a generalized back-up/cancel/delete function. This use of the "*" as an escape or command key is defined in a recent ISO/ANSI voice messaging user interface standard (see the section on standards later in this chapter).

2.1.10   Unprompted Options

Although there is a general guideline that claims that no available option should go unprompted, there may be some circumstances in which it is reasonable:

♦   Global keys. These are common global options that are not significant for a transaction and can be expected to be well known. For example, the system might refrain from prompts for "*0" for help on every menu in the user interface. Instead, prompt only when "*" is pressed, or provide an announcement at the beginning of the user dialog such as: *For help at any time, press "*0."*

♦   Hidden treasures. These are key sequences from a previous version or other vendor application provided to cater to user habits. When the application has its own key sequence to accomplish an action, e.g., "3" to delete, but users may know another key sequence from a previous version, e.g., "*3" to delete, the designer may want to allow "*3" to complete the action instead of returning an error message. However, it may be wise not to prompt for "*3" to delete, so new users don't learn obsolete keys. (This is analogous to Microsoft Excel accepting Lotus 123 key commands in its formulas, even though it has its own formula syntax.)

2.2   Call Answering

Because people bring with them rich expectations from their familiar interaction with answering machines, the designer flirts with disaster to violate the well-familiar script of answering machines:

1. recorded message in the called party's voice, saying "leave a message"

2. a record tone

3. caller leaves a message

4. hang-up terminates and saves message for subscriber

The voice mail system, however, provides a rich set of useful functions for the caller. Editing functions allow callers to delete and re-record a message they don't like, to abandon leaving a message (without leaving a blank or half-recording message after hang-up), and to navigate elsewhere in the voice mail system (e.g., leave a message for someone else in the same office, or go to the caller's own mailbox and get his or her own messages).

Figure 1 depicts a very common call answering call-flow that preserves the answering machine style of interaction while opening up the editing features. The caller who hangs up after leaving a message is confident in knowing that a message is left for the person called. The principle that hanging up should result in the system's accepting a message is very strong—also based upon answering machine behavior—so strong, in fact, that it should, perhaps, be considered as the default action whenever the user is making a recording. Certainly, when a user is recording a message to be sent by voice mail, and even when the user is recording a greeting, hanging up should trigger accepting and saving the recording.

The caller who enters a keypress is greeted with editing options. The "#" key is required by ISO/IEC 13714 as the delimiter for voice input. The "1" key is perhaps the next most common choice for this function. This is a case in which propagating the use of a standard key, the "#" key, is highly useful. The caller must know what key to press, and it's awkward to prompt for this option in the caller interface. Yet, callers have no way of knowing what brand of voice mail system is answering their calls, or whether they've encountered voice mail or an answering machine! How is the caller to guess what key, if any, will provide them with editing functions?

Prompting may be the only reliable solution until the industry accepts a standard and use of the "#" key becomes widely known among users. On one vendor's systems, a guiding prompt to this effect is inserted after subscribers' call answering greetings based on a "mailbox class-of-service" setting. For example,

Subscriber:	*Hi this is Bill. I'm not in the office right now, please leave a message after the tone.*
System:	After leaving your message, press "#" to edit your message. \<Tone\>

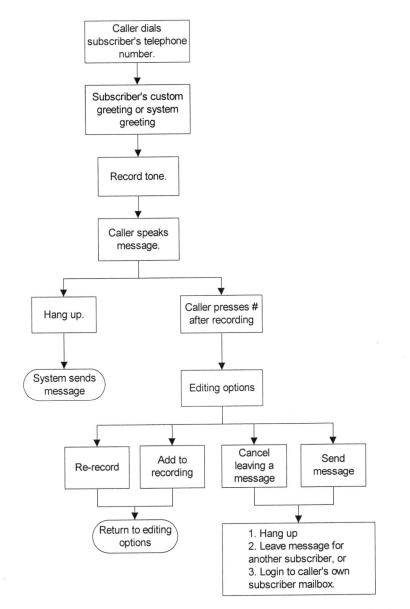

Figure 1.   A typical flow of events in a caller interface. The transaction is like that of
an answering machine unless the user enters a touch-tone (#).

Intelligent telephony agents and call-redirection systems pose an interesting problem. How disturbing is it to callers to dial someone they know, only to get an automated system trying to assert its own personality?

System:	Hi, I'm the personal agent for
Subscriber:	(recorded) *Bill*
System:	If you'd like to leave a message, press 1.
	If you'd like me to page him, press 2,
	To call him at home, press 3.

Until such systems become commonplace, we can expect them to be jarring to some users. A possible compromise is to allow subscribers the option to record the call-redirection options in their own voice, as a personalized greeting. Then, callers receive a greeting with the voice they expect, but also are offered the set of options that are available to them.

2.2.1   Call Answering Greetings

In addition to editing functions, the voice mail system offers another value to the caller interface: customized subscriber greetings that change according to relevant states. The most common of these is to have a different greeting when the called party is busy versus not answering the phone.

Subscriber	*Hi, you've reached Bill.*
greeting:	*I'm on the other line right now.*
	You can call back later, or leave a message after the tone.

This is one way to give callers an option depending upon the urgency of their communication. Various voice mail systems offer additional greetings to be played after hours, on weekends, during extended vacations, or in the case of permanent absences. The other dimension is to allow the subscriber to pre-record alternate versions of each type of greeting. The complexity of this system can quickly multiply into what can be the most complex and confusing module of a voice mail system. As a result, customized greetings also risk being one of the least-used features of a voice mail system.

2.3   The Subscriber Interface

Retrieving, manipulating, and sending messages, and administering greetings and options are all part of the subscriber user interface, which is the user-system dialog experienced by the owner of the voice mailbox. Typically, all the functions are invoked from a *main menu,* which serves as a "home base" in the dialog. A typical menu structure is outlined in Figure 2.

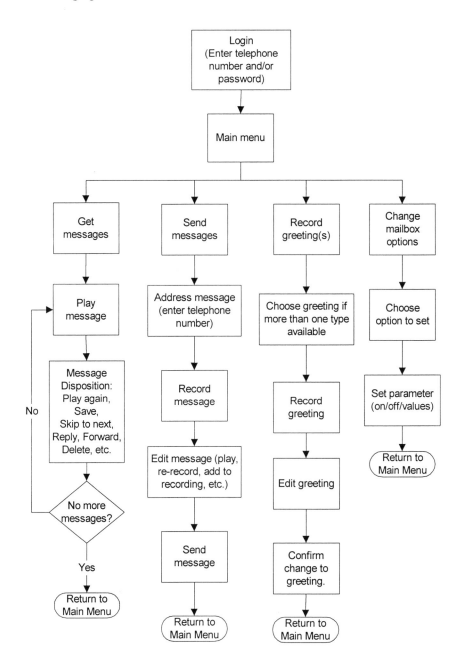

Figure 2.   A typical menu hierarchy for a voice mail subscriber user interface. The main menu is a "home base" to which the user returns after completing (or interrupting) a subtask.

2.4   Retrieving and Manipulating Messages

Playing messages is the most basic and most frequent task for voice message system subscribers. Consequently, the steps needed to play a message should be minimized. Playing messages should be the first or prominent option in a main menu. After logging in, the user should be required only to press one key, or no keys, to begin Messages are categorized into at least two types: new messages and saved messages. Messages of these two types should be kept in separate *message queues*. That is, the user should be able to choose to listen to only new messages, only saved messages, or both. New messages should be offered, or played automatically, first.

Another message category common to many systems is a "hold as new," "played but not saved," or "skipped" category. This class of messages should not be played in a separate queue. Rather, these messages are integrated with the new messages. Use of this category depends upon the default action, defined in the service, after the subscriber listens to a message. A system automatically may save a new message and move it to the saved queue if the user chooses to move to the next message. Alternatively, the system may keep the message as new if the user moves to the next message. The user must take explicit action to save the message, in order to move it to the save queue. A third design option is to save messages automatically, but provide the menu option of "keeping the message as new."

The analogy is to e-mail, where you can read new messages and keep them in your "inbox" before filing them or deleting them. This is a difficult concept for many users, and designers should consider dispensing with it if no significant user need is met by this "hold as new" class of messages.

A typical sequence for retrieving messages is as follows:

1. User selects "play messages" from the main menu.

2. The first new message is played.

3. If there are no new messages, the first saved message is played.

4. During message playback, the user can speed up, slow down, skip ahead or replay the message. Also, the user can save, delete, or skip the message if they know the command.

5. After the message finishes playing, the user hears a menu giving options for disposition of the message that includes saving, deleting, replaying, replying to, and skipping the message.

6. Once the user indicates disposition of the message by a keypress, the system plays the next message.

7. If at the end of the queue of new messages, the user is given the option of listening to saved messages.

8. At any time the user may exit the cycle of playing messages by hanging up or using a special keypress to go to the main menu or other system menu.

A default action on a message should always save the message, either saving it as a new message or as a "saved" message in a saved queue. The default action should be performed if the message is interrupted by a hang-up or if the user chooses to go back to the main menu (or other menu) by pressing a key during message playback.

When playing new messages, the oldest new message should be played first, followed by other messages in order of their arrival time, until the newest message is played. One reason for this is to play sequential messages in the order in which they happened. For example, the caller may leave a first message that explains the background of a sales deal. The same user then leaves another message an hour later, giving an update on the progress of the deal. When the user retrieves these messages, the newer message would be hard to understand without the background of the deal given in the earlier message. Chances are, the user might have to listen to both messages again to fully understand what was going on (thus doubling the time it takes to retrieve and understand the messages). If, however, the earliest message is played first, the background information is given first, and understanding follows simplistically.

2.5   Sending Messages

The promise of voice mail as more than a glorified answering machine in the network depends on the functions related to sending voice messages. Messages can be sent directly, with the subscriber entering one or more addresses and recording a voice message. Messages that already exist in a mailbox can also be sent using the reply and forward functions. If a message has been received from another subscriber, replying skips the addressing steps. Forwarding skips the recording steps, although many systems allow subscribers to attach a message of their own to forwarded messages.

A typical sequence for a subscriber sending a message is:

1. The user selects send or compose from the main menu.

2. The system begins the message-creating dialogue.

3. The user is prompted for one or more addresses.

4. The user is prompted to record the message to be sent.

5. Various post-recording options are offered.

6. The user gives input confirming that the message should be sent.

The order of addressing and message recording varies from system to system.

A little-discussed fact of voice messaging is that many users actually *don't send* messages. Unpublished AT&T survey data and actual usage data suggest that sending voice messages from the subscriber interface is, while not rare, a relatively less common voice messaging operation. (This has also been suggested from survey research by Rice and Tyler (1995).)

Although sending voice messages seems to follow a fairly simple sequence of steps, there are actually a number of usability traps. Perhaps the most significant usability problem in sending involves positive transfer from the more familiar answering machine scenario. Many users simply hang-up after they are done recording a voice message. If the system send task order is record-then-address, this means that the message can't be sent, because no recipient has been defined. Even with the address-first order, most messaging systems don't send messages on hang-up. This problem doesn't go away with extensive experience with voice messaging on different systems, as both of us can attest.

Address entry also has some usability issues. Unlike call answering, in which the user's greeting provides confirmation that the message will get to the right person, feedback for address in some voice messaging systems is simply the playback of the touch-tone digits the sender has entered. Often, this occurs without the benefit of appropriate pauses and intonation patterns that allow chunking of the numbers. On the system we use, positive acknowledgment of a successful address is marked by the absence of an error message following readback of the number. Many PBX voice mail systems help with these problems by providing "nameback," playing the recipient's name as it was recorded as a confirmation.

Readers familiar with the GOMS family of task analysis (Card, Moran, and Newell, 1980, 1983; John and Kieras, 1994) might be able to see why it may actually make sense for voice mail subscribers to leave messages using call answering rather than sending messages. Using call answering, messages are "sent" by calling the person's number, waiting for the system to answer, listening to the greeting and the record tone, speaking the message, and hanging-up. Sending messages directly from the voice mail subscriber interface involves: 1) logging into the system (which can be skipped if the subscriber is already logged in); 2) choosing send from the main menu; 3) entering the address (often the same as the telephone number) and delimiting it; 4) listening to nameback or numberback; 5) listening to the prompt for recording; 6) recording the message and delimiting it; 7) choosing among various send options (e.g., private and/or priority messages, future delivery); and, finally, 8) confirming the send with a keypress. Informal data we collected from a variety of voice mail systems while preparing this paper suggest that sending messages takes more time, involves more steps, and is more cognitively demanding than calling and leaving a call answering message.

One way around this is to encourage users to use the shortened task sequences of sending by replying and forwarding messages. Indeed, this is often a tip in corporate voice mail user guides.

2.6   Voice Messaging User Interface Standards

With the growth of voice messaging, people began to move from system to system and to have multiple mailboxes. Because the various proprietary, usually PBX-based, business voice mail systems that emerged in the 1980s had different, often arbitrary, key bindings for core tasks, reports of destructive inconsistencies surfaced. What skips ahead within a message that you're listening to on one system deletes it on another. What lets a caller skip the greeting on most systems causes the system to say "fax included" for people who call us. After a decade of rapid growth in the 1980s, formal standardization of the touch-tone user interface to voice messaging became likely due to strong support for a standard among service provider, equipment manufacturer, and user group market segments.

Callers' user requirements were also a standardization driver. Callers can be thought of as *involuntary* users of systems who encounter someone's voice messaging system and leave messages. They often don't know and almost always don't care what kind of system they have reached. Many callers use voice recording but no other features when leaving messages, but there are a half-dozen or so features that some callers do use. These include: 1) skipping greetings; 2) listening to, re-recording, or deleting messages they are leaving; 3) terminating their voice message; 4) trying to reach a live person; 5) getting help; 6) forcing the system to disconnect; 7) setting privacy or urgency options for a message, and 8) logging in to their own mailbox after leaving a message. Industry and standards group participants believed that defining a common user interface for a few of these features could have large benefits to callers.

Voice messaging user interface standards work began in industry associations and moved to formal standards groups, ANSI in the U. S. and ISO internationally. There was broad organizational participation in this work, with many issues of *what* should be included in a standard and *how* the standard should implement common features, leading to contention and compromise. Once the international work began, this voice messaging user interface standard progressed quickly and was published in March of 1995 as ISO/IEC 13714 (ISO/IEC JTC1, 1995). Identical text with an ANSI cover and introductory material was published as ANSI/ISO 13714 in May of 1996. This standard is titled "User interface to telephone-based services: voice messaging applications."

The standard requires very little in terms of voice messaging features or their user interface:

- ◆ A record tone is required. One is recommended but no specific tone is mandated.

- ◆ Systems must offer context-sensitive help, which must be accessible after the user inputs a "*0" or a "0" alone.

- ◆ The user must be able to dial-through system prompts, announcements, messages, etc., except for a few cases specifically called out in the standard.

A user interface standard can't really require *features* of a voice messaging system. Systems typically have record tones, context-sensitive help, and dial-through. More important, the standard defines required implementations for about a dozen core voice messaging activities.

The wording in standards differentiate required features from those which, if implemented, must have a specific user interface. In practice, since almost all of the features in this category are found on nearly all current systems, this is a strong requirement. The gist is that "if you provide capability x, it must by implemented as y, for your system to be conforming." These required implementations define a few basics of how users deal with messages, prompts, voice input, data, menus, and navigation. Depending on how the counting gets done, the standard covers about a dozen common voice messaging tasks. These are summarized in Table 1. ISO/IEC 13714 also collects and presents design guidelines and recommendations that reflect best industry practices for voice messaging user interface design.

In voice messaging, *callers* are people who encounter someone's system and leave messages. Although this can be as simple as dialing, waiting for an answer and a record tone, then leaving a message, callers actually often need to be able to do more. It's useful to callers to be able to skip greetings. ISO/IEC 13714 promotes interoperability of systems for caller control of some basic functions. When systems conform, there will be fewer cases in which what gets help on one system cancels message recording and causes a hang up on another, or in which what skips through the greeting to the record tone on one system prompts the caller to login on another. Although in the big picture of voice messaging user/system interaction these caller features may not be used that often, the strong standards case for callers contributed to a relatively large number of caller features being included. Several of the required implementations for callers came from *de facto* industry standards, at least in the U.S., on the use of "#" and "0."

Table 1.   ISO/IEC 13714 standardized features.

Feature		Required implementation
Caller interface		
1	Skip mailbox greeting	#
2	Delimit message recording	#
3	Delimit DTMF Input	#
4	Get system help	*0
5	Force system disconnect	*9
6	Has a record tone	Must be provided
7	Attendant/coverage	0
8	Dial-through	Must be provided
Subscriber interface		
	(all above, plus)	
9	Return to main menu	*7
10	Main menu: Listen	1
11	Main menu: Send	2
12	Listen menu: (Re)Play	1
13	Listen menu: Delete	3
14	Listen menu: Skip message	#
15	Send menu: Send message	4
16	Backup one 1 level or step	**
17	Cancel voice or DTMF input	**
18	Start voice recordings	#

For *subscribers*, most of the standardized caller interface features apply, and the identical key bindings are required implementations. Most of what is additionally required for the subscriber interface are menu key bindings for the most common things people do when they access their voice mailboxes, particularly listening to and saving or deleting messages. If systems conformed, there would be fewer cases in which what skips ahead on someone's work voice messaging system deletes it on their home system, or in which what replays a message on one system deletes it on another.

With the publication of this formal user interface standard, voice messaging became, perhaps, the first application to have a comprehensive user interface standard. It is still too early to tell if this work will produce the user and industry benefits promised. There has been implementation of various parts of the standard by some vendors and some service providers, but legacy code, products, and even users have been highly resistant to change.

2.7 Alternative Approaches to Traditional Touch-tone Design

Voice mail systems, as well as most other IVR systems, use a fairly standard and ubiquitous dialog technique: a series of prompts with options, i.e., a "menu" format. We have assumed that the voice mail application follows this dialog structure in our discussion thus far. But what about some different, innovative approaches to touch-tone user-system interaction? For example, in the IBM Speech Filing System (an early IVR application incorporating voice mail and voice reminder functionality), Gould and Boies (1983) designed what was basically a touch-tone command system. The user entered two-key touch-tone sequences without being prompted for those sequences. There were no prompts, but there was feedback given as to which command was invoked. So, for example, to compose and send a message, the user would press "*R" to record, the system responded with "Record", the user recorded the message, then pressed "*T" to send (transmit) the message. The system responded by saying "Transmit"; then the user entered the telephone number of the recipient, and so on. Later, Gould and Boies (1983) switched to the modern prompted menu format. Although they noted that the menu format allowed users to use their system without instruction, the speed of interaction suffered. Over the last decade, the advantage of the menu approach, in allowing voice mail to be used without detailed instruction, has overwhelmed any advantage in speed obtainable from a command approach. Using type-ahead, expert users can also simulate a command approach. However, a menu structure optimized for menu use is not optimally efficient for a command approach. For systems that have a large percentage of expert users, like voice mail, it might be worthwhile to revisit the command approach. The best technique might be to incorporate elements of commands within a menu system, to allow both novice and expert use. For example, a set of "*-digit" command sequences could be active throughout the interface to invoke a set of commonly used functions.

Recently, Resnick and Virzi (1995) proposed a number of other dialogue formats for touch-tone systems as possible alternatives to the common menu design (see also Resnick and Virzi, 1992). They proposed a variety of different techniques, some of which would be more applicable to IVR systems that have simple transactions and do not have expert users. None of their schemes were extensively tested with users, but certainly different approaches may be worth exploring as novel solutions to certain interactions. For example, Resnick and Virzi proposed one scheme they call a "rejection menu dialog," which is a kind of "twenty-questions" approach to menu choice. Using voice mail functions, one rejection menu dialogue might be as follows:

System:	If you'd like to get messages, please wait, otherwise, press #.
User:	#
System:	Send a message. Wait to select or press # to reject.
User:	#
System:	Change your greeting. Wait to select or press # to reject. [etc.]

Obviously, this scheme does not transfer well to the main dialogue in a voicemail system. However, it may be more appropriate for simple two- or three-choice options, which do occur at many points in voice mail systems. (The rejection dialogue might also be cast as a sequence of yes/no or two-choice option questions.)

Overall, alternative dialogue techniques such as commands or rejection menus may be useful to explore in voice mail systems as *supplements* to the main design rather than substitutes for the standard menu approach.

3. AUTOMATIC SPEECH RECOGNITION AND VOICE MAIL

The next technology for interaction with computers over the phone is speech recognition by machine. Voice commands can supplement or replace the use of the touch-tone keypad in controlling a voice mail application. Elsewhere in this book (Chapters 7 and 13) the problems and challenges associated with using speech recognition in IVR have been discussed, and certainly these same considerations apply to voice mail applications.

In addition to usability considerations for automatic speech recognition systems in general (e.g., Karis and Dobroth, 1995), voice mail presents two problems that are not often significant in other telephone applications:

1. Voice mail systems are functionally complex and thus require complex vocabularies.

2. The system must distinguish between times when the user is talking to the system and when the user is recording a message.

Functional complexity places a burden both on the user and on the speech recognizer. The user must remember a long list of commands, which means either mastering the keywords of a system, or paying close attention to menus that exhaustively list commands and their functions. Large vocabularies may reduce the effectiveness and accuracy of the speech recognition system. Depending on the technology, the system may perform better with phrases than single word commands (e.g., "Help me out" rather than just "help"), but this imposes an even further onerous memory burden on the user. The solution to this is more flexible natural language systems (cf., Chapter 2) that accept many variations on natural phrases.

A fruitful technique for dealing with experts versus novice users of a voice command system is used by Brems, Rabin, and Waggett (1995). In essence, Brems et al. suggest:

1. Choose commands that are transparently descriptive of their function.

2. Choose a technology that allows barge-in of voice commands.

3. Prompt only for the commands first.

4. Pause for experts to respond. This allows expert users who know the commands to respond right away without having to listen to a long explanation.

5. Follow the pause with a more detailed menu for novice users.

For example, applying the Brems et al. suggestions to voice mail, a menu might be presented this way:

System:	[System plays message]
System:	Please say delete, reply, call back, forward, play again, or next message.
	\<pause\>
System:	To erase this message, say delete, to answer this message by voice mail, say reply, …

The ability to skip over long prompt explanations is a crucial element for giving a speech recognition user interface an advantage over the touch-tone alternative. Franzke, Marx, Roberts, and Engelbeck (1993) found that users of a speech recognition user interface to a voice mail system completed tasks faster than users of a touch-tone version of the same system. The main reason for the advantage was that speech users learned a set of simple commands and readily interrupted the prompts, whereas touch-tone users tended to listen to prompts. (Note, though, that these were novice users observed only through a short experiment, and that Fay (1993) and Karis (1997) found the opposite results with different applications.)

Also a challenge both to usability and technology is the need to distinguish when the user is addressing a command to the system or recording a message. To illustrate, imagine a system accepts the command "erase" while the user is making a recording. If the word erase appears in the message content (e.g., "we'd like to *erase* our third quarter losses."), the message is erroneously deleted.

This restricts the user to the use of a single keyword that catches the attention of the recognizer. (Note that this is a loss of functionality relative to a touch-tone user interface, which can allow touch-tone recording control during message recording.) The speech recognizer will monitor for one keyword or phrase such as "Ready", "End recording", "Computer!" or "Wake up." There are two technical problems with this. One is the burden on the processor, which must continuously recognize the speech stream in order to monitor for the wake-up command, and the other is finding a word which makes sense to users but which is unlikely to be used in conversation. (Compare, for example, the Wildfire call management system, which monitors for the word "wildfire" during normal telephone conversations, presumably with the hope that wildfires are not a common subject of business discourse.)

Of course, the drawback here is that the user must remember to use the wake-up keyword without immediate prompting. Depending upon the capabilities of the technology used, it may be more practical to use a touch-tone command to terminate a recording. But if the ASR system is being targeted to users of rotary telephones, this poses a problem. Recordings can terminate via a time-out period, but this could quickly become tedious to users, because the time-out interval must be longer than pauses found in normal conversation.

Fay (1993) and Karis (1997) have found that users may actually prefer touch-tone-based systems to voice command systems, due at least in part to the limitation of speech recognition technology. From early applications of the technology, human factors experts have noted (e.g., McCauley, 1984) that users may not tolerate recognition errors associated with ASR unless there is a distinct advantage to be obtained from using voice control. Most often cited is the ability to control a device in a "hands-free" mode. In the domain of voice mail, the need for hands-free control for accessing voice mail from a cellular car phone is well noted. This is a natural usability advantage that can be exploited to increase the acceptance of ASR technology.

4.   UNIFIED MESSAGING AND MULTIMEDIA MAIL

Ironically, voice mail began as a sort of telephone analog of the terminal-based electronic mail systems of a decade ago. Now, the similarities between voice mail and e-mail are pushing the technologies to merge into *unified messaging*. Unified messaging, in its ultimate form, is the combination of voice and electronic messages and faxes into one single mailbox, which can be accessed through an IVR system over the telephone or through a mail client on a computer. A review of recent unified messaging platforms and service can be found in Kosiur (1998).

4.1   Fax Messaging

This trend certainly began when voice mail systems began to add a premium feature - the ability to intercept and electronically store faxes with the user's voice messages. Faxes can be received by the voice mail system in two ways:

♦   Faxes are sent to the user's regular telephone number. When the voice mail system answers the call, it detects the fax tone, receives the fax electronically, and stores the fax as a special class of message.

♦   Faxes are sent to a second number, a fax number. The voice mail system again stores incoming faxes electronically. The fax messages then are transferred to the user's voice mailbox.

Alternative configurations of these fax features are also offered by vendors to solve various office problems. So, for example, a busy fax machine can be given a voice mailbox. When a fax comes in while the machine is already receiving a fax, the fax machine's voice mail picks up and stores the new, incoming fax for later printing.

Fax call answering presents an enormous usability problem that the industry has yet to solve effectively. When the system uses the regular voice telephone number to receive fax messages, things work well as long as the user doesn't answer their phone. However, it is not possible to detect a fax call before an incoming call is picked up. Thus, subscribers inevitably must tolerate getting a fax tone sometimes when they answer a call.

A second solution, using a separate fax number, thus may be better from a usability point of view. (The user's callers must know two numbers to call for fax versus voice, but that is a very familiar convention.) Unfortunately, this solution is expensive, for there must be either a separate extension for every person in the organization (thus doubling the needed capacity of the phone system) or the fax mailbox must be shared among users. The shared fax mailbox is also expensive in terms of human time, for a person must manually view the faxes and then forward them to the appropriate user mailboxes. This remains a challenge for fax and unified messaging.

4.2   Viewing Voice Mail

The next voice mail feature, which has been offered as a separate component, is the ability to view a list of voice mail messages on a personal computer and to listen to them (provided the user's computer has sound capability). At least part of the rationale for offering the ability to view voice mail headers is based upon a human limitation: it is difficult to manipulate lists of items using the auditory medium. Using the telephone, the user is forced to search messages one-by-one in a fixed sequence. The user may also need to rely on memory to recall the relative order of messages in the list. The user might be able to skip over messages but still must go through them in serial order. Systems can be designed to let users jump to a message using a "message number," for example:

System:	You have twenty saved messages.
	What message do you want to hear?
	\<pause\> enter the number of the message.
User Enters:	*17*
System:	Message 17. July 7, 1998. ...

Obviously, this poses an impractical memory burden on the user. Can we expect users to know the order in which they saved their messages? Another option is to provide audio skimming and search capability through speech processing technologies (Marx and Schmandt, 1996; Roy and Schmandt, 1996). These are promising techniques that could be applied to voice mail retrieval, but they are still in the research stage.

This difficulty is eliminated if the user is able to see a visual list of his or her messages. The user can scan all the messages in a visual list and does not need to recall the order of a message in a list. It is then a simple matter to choose the message to manage (play, delete, reply to, forward, or file the message) from a complete list. The message could be manipulated from the computer user interface or from the telephone IVR system. Fusco and Gattuso (1991) found that not only did users prefer to use a visual interface when using the more complex features of voice mail, but also users found the voice user interface less confusing after using the visual user interface.

This same capability also has been offered on display phones, where a list of voice message headers can appear on a large display on the telephone. Buttons next to the display, called "soft keys", allow users to choose a message to manage. Such a system is most usable if there is a tight and elegant integration between the telephony user interface and the display.

For example, the user is presented with a list of messages, one per line, with a soft key to the right of each message header. Soft keys at the bottom of the display are labeled "delete," "reply," and so forth. While the user holds the handset and presses a message button, the message is played. After the message is played, a touch-tone menu is played. The user can press the "delete" soft key over the menu in order to quickly delete the message just played. The response should be immediate and real time with the IVR system.

Soft key systems have not proliferated, however, particularly in the face of personal computers. Because a user is likely to have a personal computer, particularly in the business environment, and that computer is likely to be close to the telephone, computer-based applications are more practical.

Having a separate application for voice mail may seem redundant, since the major functions are identical to an e-mail client program. Hence, unified messaging seems likely to supplant separate applications that display voice mail messages.

4.3   Listening to E-mail

It has long been possible to listen to electronic mail over the telephone using text-to-speech (TTS) technology. Indeed, AT&T Mail, among others, had offered MailTALK℠ for its customers eight years ago. Perhaps because of the greater use of e-mail over the last few years (associated with the prominence of the Internet), there has been a resurgence of interest in reading e-mail over the telephone.

There are four notable usability considerations with e-mail over the telephone:

1. *The acceptability of TTS technology.* Even with the technology of a few years ago, text-to-speech is highly intelligible. Users are nearly as accurate as with natural speech in understanding TTS (Paris, Gilson, Thomas, and Silver, 1995; Pisoni, Manous, and Dedina, 1987; Syrdal, 1994). Listening to TTS may require more mental processing, and can be more exhausting than natural speech to listen to in large blocks. The real user impediment to TTS, however, is that users simply dislike the sound of TTS regardless of how intelligible it is (Cowley and Jones, 1992).

2. *Transliteration of textual conventions to voice.* E-mail messages are full of textual conventions that are visual in nature, and do not translate to speech directly. Smiley faces (e.g., ;-)) are an obvious example, but even more ubiquitous are lines of dashes or stars used as rules or demarkations. The system must be suppressed in some manner to prevent it saying something like the word "dash" eighty times in the middle of a message. E-mail messages also have headers with detailed routing information. Reading of those headers should also be suppressed (Schmandt, 1995). The common way this suppression is

accomplished is by pre-processing the e-mail text before it is given to the TTS processor. A significant improvement to usability can be obtained with more sophisticated pre-processing.

3. *The sheer volume of e-mail messages for some users.* For many heavy, or even moderate, users, the number and frequency of e-mail messages far outstrips the number of voice mail messages one might typically expect to receive. Where it might make sense to light a message light to signal the arrival of voice mail, chances are it makes no sense to light a message light on a telephone when the user receives e-mail, since that light would be on constantly. The high volume of e-mail poses problems for inventory of messages, user navigation among messages, filtering through relevant as opposed to junk messages, and time spent listening to messages.

4. *Navigation among e-mail messages.* Earlier, we mentioned the difficulty of randomly accessing voice mail messages when using a telephone-based user interface. Messages must be accessed sequentially. The same problem is just as true when moving among e-mail messages over the telephone.

An application that reads e-mail is very much like a voice mail application in its structure and functionality. The next logical step is to combine e-mail and voice mail into the same application.

4.4   Putting it All Together

Unified messaging puts all of the previously described features together, with the merging of all message types into one common mailbox. Voice, e-mail, and fax messages are stored in the same mailbox. A computer program, most likely an e-mail client, can access the mailbox. Voice, e-mail, and fax messages all are shown in the message list, and voice messages can be played by the computer (provided proper hardware and software are present, of course). In addition, there is a telephone application, the extension of a voice mail application, that plays voice messages, reads e-mail messages by TTS, and lists fax messages and forwards them for printing.

In addition to all the usability considerations for the separate elements discussed previously, unified messaging poses the additional usability question: how much access to messages do users really want?

This is particularly relevant to telephone access to the unified message mailbox. Access to three kinds of messages means triple the message queues to listen through. Separating message types must be done, as e-mail will overwhelm voice and fax in volume. Some users may dislike e-mail TTS. Extra menus must be added to a voice mail application framework in order to navigate among the message types.

Perhaps, as users adapt to the functions offered by unified messaging, it is best to provide a wealth of user customization options for the telephone application. For example, allow users to suppress access to any of the three message types. Some users could configure their unified voice mailbox to get just their voice mail, making it like their more familiar voice mail. Others could get only voice and fax messages; still other users could choose to get all three types. Allow configuration of e-mail access, so that a user who dislikes TTS can skip over the reading of e-mail message bodies, or have an option that only reads e-mail headers.

4.5   Mixed Media

The unified mailbox is really a multimedia mailbox: visual and auditory messages are mixed in the user interface. The user can use a visual (computer) user interface or an audio (telephone) user interface to work with those voice and visual messages. A simple extension of such a design is to allow mixing of media in a single message. Thus, the user can receive a fax that is prefaced by a voice annotation.

The next opportunity that presents itself is the freedom to respond to messages in any mode. Thus, the user receives a voice mail message through the computer user interface, and decides to respond with an e-mail message. The recipient of the response, if at their phone, can still listen to the message, since the e-mail can be read by text-to-speech.

How valuable is this? This is the major challenge to user interface designers for the next generation of messaging: how do people want to get their messages, and do they need the freedom of mixed media at the cost of increasing complexity of the user interface?

It has often been observed informally that there are individual differences in preference for visual versus auditory communication. Some users prefer to use e-mail, others prefer to use voice mail, and one learns in an organization the optimal method to contact certain individuals. The phenomenon is ripe for systematic study. Multimedia messaging allows users to communicate in their own preferred media rather than in that of their correspondent. But is this really a problem? Will users simply tend to reply using the mode by which they were first contacted?

Of course, the other variety of multimedia mail is call answering for videotelephones. If videotelephony ever becomes common, then videotelephony mail will follow. The video mail system will pose new user interface challenges based upon the capabilities of the videotelephone device (be it a telephone and/or a computer). Video mail also then will become another medium to manage in a unified mailbox. However, the existence of video mail as a consumer service is dependent upon the regular use of person-to-person video calling (as opposed to arranged video conferences), and the history of attempts to introduce person-to-person video calls is not encouraging (Blanchard and Angiolillo, 1994; Noll, 1992).

REFERENCES

American National Standards Institute (ANSI). (1996, May). *ANSI/ISO 13714: User interface to telephone-based services: voice messaging applications.* New York: Author.

Blanchard, H. E., and Angiolillo, J. S. (1994). Visual displays in communications: A review of effects on human performance and preference. *SID International Symposium Digest of Technical Papers, 25*, 375-378.

Brems, D. J., Rabin, M. D., and Waggett, J. L. (1995). Using natural language conventions in the user interface design of automatic speech recognition systems. *Human Factors, 37*(2), 265-282.

Card, S. K., Moran, T. P., and Newell, A. (1983). *The psychology of human computer interaction*. Hillsdale, NJ: Erlbaum.

Card, S. K., Moran, T. P., and Newell, A. (1980). The keystroke-level model for user performance with interactive systems. *Communications of the ACM, 23*, 396-410.

Clark, H. H., and Haviland, S. E. (1977). Comprehension and the given-new contract. In R. O. Freedle (Ed.), *Discourse production and comprehension* (pp. 1-40). Norwood, N.J.: Ablex.

Cowley, C. K., and Jones, D. M. (1992). Synthesized or digitized? A guide to the use of computer speech. *Applied Ergonomics, 23*(3), 172-176.

Englebeck, G., and Roberts, T. (1990). The effects of several voice-menu characteristics on menu-selection performance. In *Proceedings of the Bellcore Symposium on User Centered Design* (pp. 50-63) (Bellcore Special Report SR-STS-001658). Red Bank, NJ: Bell Communications Research.

Fay, D. (1993). Interfaces to automated telephone services: Do users prefer touchtone or automatic speech recognition? In *Proceedings of the 14th International Symposium on Human Factors in Telecommunications* (pp. 339-349). Darmstadt, Germany: R. v. Decker's Verlag.

Franzke, M., Marx, A. N., Roberts, R. L., and Engelbeck, G. E. (1993). Is speech recognition usable? An exploration of the usability of a speech-based voice mail interface. *SIGCHI Bulletin, 25*(3), 49-51.

Fusco, M., and Gattuso, N. (1991, September). *An assessment of a visual voice mail interface*. Poster presented at the 35th Annual Meeting of the Human Factors Society, San Francisco, CA.

Gould, J. D., and Boies, S. J. (1983). Human factors challenges in creating a principal support office system—The speech filing system approach. *ACM Transactions on Office Information Systems, 1*(4), 273-298.

Halstead-Nussloch, R. (1989). The design of phone-based interfaces for consumers. *Proceedings of CHI-89: Human Factors in Computing Systems* (pp. 347-352). New York: ACM.

John, B. E., and Kieras, D. E. (1994). *The GOMS family of analysis techniques: Tools for design and evaluation.* (Carnegie Mellon University Human Computer Interaction Institute Technical Report 94-106). Pittsburgh, PA: Carnegie Mellon University.

ISO/IEC JTC1 (1995). *ISO/IEC 13714: User interface to telephone-based services: Voice messaging applications.* Geneva: Author.

Karis, D. (1997). Speech recognition systems: Performance, preference, and design. In *Proceedings of the 16th International Symposium on Human Factors in Telecommunications.* Oslo, Norway.

Karis, D., and Dobroth, K. M. (1995). Psychological and human factors issues in the design of speech recognition systems. In A. Syrdal, R. Bennett, and S. Greenspan (Eds.), *Applied speech technology* (pp.359-388). Boca Raton: CRC Press.

Kosiur, D. (1998). Making life simpler with a universal mailbox. *Internet Computing, 3*(9), 87-89.

Lee, E., and MacGregor, J. (1985). Minimizing users' search time in menu-retrieval systems. *Human Factors, 27*(2), 157-162.

Leopold, F. F., and van Nes, F. L. (1984). Control of voice mail systems by voice commands. *IPO Annual Progress Report, 19*, 118-122.

Marics, M. A., and Engelbeck, G. (1998). Designing voice menu applications for telephones. In M.G. Helander, T.K. Landauer, and P. Prabhu (Eds.), *Handbook of human-omputer nteraction* (2nd ed.) (pp..1085-1102). New York: Elsevier.

Marx, M., and Schmandt, C. (1996). MailCall: Message presentation and navigation in a non-visual environment. *Proceedings of CHI-96: Human Factors in Computing Systems* (pp. 165-172). New York: ACM.

McCauley, M. E. (1984). Human factors in voice technology. In F. A. Muckler (Ed.), *Human factors review: 1984* (pp. 131-166). Santa Monica, CA: Human Factors Society.

Noll, A. M. (1992). Anatomy of a failure: Picturephone revisited. *Telecommunications Policy, 18*, 307-316.

Norman, K. L. (1991). *The psychology of menu selection: Designing cognitive control at the human/computer interface*. Norwood, NJ: Ablex.

Paris, C. R., Gilson, R. D., Thomas, M. H., and Silver, N . C. (1995). Effect of synthetic voice intelligibility on speech comprehension. *Human Factors, 37*(2), 335-340.

Pisoni, D. B., Manous, L. M., and Dedina, J. J. (1987). Comprehension of natural and synthetic speech: Effects of predictability on the verification of sentences controlled for intelligibility. *Computer Speech and Language, 2*, 303-320.

Resnick, P., and Virzi, R. A. (1992). Skip and scan: Cleaning up telephone interfaces. In *Proceedings of CHI-92: Human Factors in Computing Systems* (pp. 419-426). New York: ACM.

Resnick, P., and Virzi, R. A. (1995). Relief from the audio interface blues: Expanding the spectrum of menu, list, and from styles. *ACM Transactions on Computer-Human Interaction, 2*(2), 145-176.

Rice, R. E., and Tyler, J. (1995). Individual and organizational influences on voice mail use and evaluation. *Behaviour & Information Technology, 14*(6), 329-141.

Roy, D. K., & and Schmandt, C. (1996). NewsComm: A hand-held interface for interactive access to to structured audio. *Proceedings of CHI-96: Human Factors in Computing Systems* (pp. 173-180). New York: ACM.

Schmandt, C. (1995). Voiced mail: Speech synthesis of electronic mail. In A. Syrdal, R. Bennett, and S. Greenspan (Eds.), *Applied Speech Technology* (pp. 389-402). Boca Raton: CRC Press.

Schumacher, R. M., Jr., Hardzinski, M. L., and Schwartz, A. L. (1995). Increasing the usability of interactive voice response systems: Research and guidelines for phone-based interfaces. *Human Factors, 37*(2), 251-264.

Syrdal, A. K. (1994). Development of a female voice for a concatenative text-to-speech synthesis system. *Current Topics in Acoustic Research, 1*, 169-181.

13 A NEW DESIGN FRAMEWORK FOR COMPUTER-TELEPHONY INTEGRATION (CTI)

Martha J. Lindeman

Users First, Inc., Columbus, OH USA

Key words: Computer, Telephone, Computer-Telephony Integration (CTI), System Design, System Development, User Interface, Voice Response, Customer Service, Automated Attendant, Agent, Interactive Voice Response (IVR), Voice-Response Unit (VRU)

Abstract: The need for computer-telephony integration (CTI) has expanded far beyond simply displaying information about an inbound call to a customer-service representative. This expansion has increased the number and types of problems experienced by developers and callers who interact with CTI systems. Some of these problems can be identified and prevented by using a design framework with the dimensions of "time" and "complexity". The time dimension applies to the five phases of a call's life cycle: (a) pre-call issues such as attributes of the caller and the caller's environment, (b) call routing, (c) task interactions during the call, (d) post-call activities, and (e) historical records and statistics. Thus the model covers both stand-alone calls (e.g., to place an order) and calls that are related to previous calls (e.g., to change an order). The complexity dimension includes three types of complexity for inbound calls: (a) technology, (b) request and (c) people. Blending outbound calls into the call mix adds another type of complexity. Failing to understand the issues and variables for either time or complexity can cause problems for developers of a system and for callers after the system has been implemented.

1. WHAT IS CTI?

Very frequently a telephone call in today's world is answered by a machine. Sometimes this technology is called "voice mail" or "automated attendant", or even technical jargon such as VRU (voice-response unit) or IVR (interactive voice response). When computers are used to display information about a caller to a customer-service representative, this has been called computer-telephony integration or CTI.

CTI now has grown far beyond this narrow definition of use to include functions such as answering calls, routing calls, and Internet access. Thus a "call center" is now more appropriately a "contact center" that includes multiple communication types. The design framework in this chapter applies to all types of "contacts", but the term "call" is still used for ease of reading.

It can be difficult to integrate various CTI technological components into a successful system, and failure to do so can cause human factors problems for callers, users, and system developers. The cumulative effect of even trivial problems can strongly decrease callers' satisfaction with a system.

CTI calls can be organized and then analyzed by mapping three sources of complexity (technology, requests, and callers) across the phases of a call's life cycle. Each life cycle event can be mapped on a timeline with five phases, ranging from before the call is made to when all historical records are completed for the call. The mapping of life cycle phases to sources of complexity create an expanded matrix with rows similar to those in Table 1.

The "technological complexity" of a call is determined by the number and types of technological events that occur during a call, such as answer, automated request for caller input, database query and retrieval, transfer to another telephone extension with simultaneous transfer of visual information, etc. (the first blank row in Table 1).

In addition to technological complexity, human factors problems in CTI systems can be caused by "request (i.e., task) complexity" and "caller complexity". Analyzing a CTI system using a matrix framework based on Table 1 can yield faster and easier system design and a more successful implementation. The rest of this chapter describes sample issues for cells in the design matrix using examples drawn from unidentified real-world systems.

Table 1.   A matrix framework for discovering potential problems in the design of a computer-telephony integration (CTI) system.

Sources of Complexity	Call Life Cycle Phases (Time Dimension)				
	Pre-Call Considerations	Call Routing	Task Interactions	Post-Call Activities	Historical Records
Technology					
Request type					
Caller type					

2.   THE "CALL PHASE" TIME DIMENSION

Every event associated with a call can be mapped on a timeline with five phases, ranging from before the call is made to when all historical records are completed for the call. The five phases of a call's life cycle are listed below:

- Pre-call issues (e.g., attributes of the caller and the caller's environment)

- Call routing (e.g., data-based routing and transfers)

- Task interactions (e.g., identifying the item being ordered)

- Post-call activities (e.g., research and call-backs, order fulfillment)

- Historical records and statistics (e.g., contact logs, abandonments)

The **pre-call** phase begins when the caller decides to make a call and ends when the call is answered. Pre-call issues for a CTI system typically include attributes of the caller population (age, technical sophistication, emotional state, etc.) and callers' environments (home, workplace, car, etc.).   Pre-call issues also include circumstances such as time of the call (e.g., after-hours or holidays) and hardware used to "dial" the call (e.g., by speech or telephone keypad).

The **call-routing** phase begins when the call is answered, whether by human or by machine, and ends when the caller begins to input information necessary to carry out the caller's request (e.g., an account number). The first entry of task information designates the beginning of the **task-interactions** phase, which ends when the caller hangs up or is released from the call. Sometimes it is difficult to differentiate between call routing and task interactions, particularly when a call is handled by both automated and human agents. However, this can be an important part of unifying the activities of automated and human agents and thus presenting a well-designed user interface to the caller.

Post-call activities, sometimes called "workflow", refer to the number and/or complexity of the processing steps that must be done after the caller's request has been received and entered. For example, the human agent answering a call may have to do further research and then return the call. In other cases, the billing, shipping, and inventory departments must process an order after it has been received and entered. For each type of request, it is necessary to identify and describe: 1) the entities that must process the request, and 2) the information and actions required at each point in the process.

Storing **historical** information for retrieval and analysis may be one of the least understood but most valuable aspects of CTI. Historical information includes data about call processing, products and services, customer-relationship management and other appropriate information.

Identifying potential problems in one cell of the matrix may help identify potential problems in other cells. For example, identifying a "caller type", such as elderly people, may also yield a "method of input" issue in that elderly callers are more likely to use a rotary telephone.

3.   SOURCES OF COMPLEXITY

Every user interface problem experienced by people who interact with CTI systems can be traced to one or more sources of complexity.   There are three primary sources of complexity: 1) technology, 2) callers, and 3) requests.

3.1   Technology

Technological complexity refers to the various hardware and software elements that are integrated together within a user interface.   For example, a technological event such as a transfer between two automated systems may be 1) imperceptible to the caller, 2) indicated by a sound such as a telephone ringing, or 3) announced by a verbal prompt.

The probability that a caller will experience some type of technological problem increases as the technological complexity increases.   The critical task from a human factors perspective is to hide the technological complexity from the caller or user as much as possible, unless there is a specific gain from making the complexity obvious to the caller.   For example, technological complexity is made obvious when callers are told that they are going to be transferred and to ignore any pops or clicks that they might hear.   Instructions such as these can help prevent callers from hanging up (abandoning) calls before transfers are completed.

3.2   Requests

Request complexity refers to the complexity of task-related information that must be known or provided by the caller.   For example, a very simple request would be to ask how late a store will be open.   The caller does not have to know anything other than the telephone number of the store.   A more complex request would be when a caller has to identify an account and one or more item numbers to place an order. Finally, a very complex request would be when a caller is asking for technical help with a computer problem that the help-desk has never seen before.   In this case, the caller would have to provide numerous types of information to allow the help-desk agent to diagnose and solve the problem.

3.3   Callers

Caller complexity can happen in a variety of ways.   For example: 1) the caller population includes groups of people with significantly different attributes; 2) an individual caller has a complex relationship with the company being called; and 3) changes over time in a caller's attributes, such as changes in marital status and dependents, can affect caller complexity (e.g., for financial records).

If the caller population includes people unfamiliar with automated systems and other people who are technologically sophisticated, then the user interface should be designed to deal with caller complexity. For example, many systems transfer callers directly to a human agent if the callers do not enter any information.

Another example of caller complexity is when an individual caller has a complex customer relationship with the company.   For example, I know someone with a home office and five telephone lines spread across three bills. The local phone company is not set up to deal with this type of complexity, and consequently this person has received multiple telemarketing calls offering an additional line that could not be added as offered. After calling the telephone company back, the person found that adding a sixth line requires a special trip by a telephone company engineer. Obviously, this is not good customer-relations management.

In many situations, caller complexity typically shows up as a failure to differentiate among the terms "customer," "account," and "caller." It is often useful to differentiate among these terms in both the database and the user interface to decrease problems caused by this type of complexity.

4.   "CALL PHASE/COMPLEXITY" EXAMPLES

Any CTI user interface is easier to design when the interactions of call phase and sources of complexity are understood. The following paragraphs contain examples for each cell in the simple matrix in Table 1. These general examples were chosen to help readers understand how to expand and complete the design matrix framework for their particular CTI systems.

4.1   Technology/Pre-call Example

Technology/pre-call issues typically involve the hardware and/or software at the point of call origination. For example, identifying where callers may be when they place the calls becomes a critical pre-call issue when automatic number identification (ANI) is used by the CTI system. For example, what proportion of callers calling from work will be automatically recognized? If none, why not, particularly if callers' work numbers can be entered into the database? What about calls that do not carry any ANI information with them?

4.2   Technology/Call-Routing Example

For a technology/call-routing example, consider a CTI system that transfers some callers to another system that controls the hold queue for a particular group of customer-service representatives. The technology used in the hold-queue system decreases the number of callers allowed in the queue as the length of wait-time increases. Thus when the wait-time exceeds the threshold, callers transferred to the hold-queue system hear a circuit-busy signal and are then disconnected. This is extremely frustrating, and it decreases 1) callers' perceptions of the service, and 2) the morale of the customer-service representatives who must deal with the callers' anger when they finally do get through the system to a human agent.

4.3   Technology/Task-Interactions Example

A technology/task-interactions example is when the available hardware and/or software does not match the information to be entered by the caller. For example, consider what happens when a caller is asked to enter a text string using a telephone keypad - the user interface has to include special instructions for how to enter "Q" and "Z" if they will occur in the text. In this case, speech recognition may be preferable as the primary mode of input with appropriate means of error correction and recovery.

4.4   Technology/Post-call Example

A major technology/post-call issue is whether or not to use automated workflow. For example, the technology/post-call cell for order entry might indicate that the CTI system automatically: 1) sends appropriate billing information to the billing system, and 2) generates and sends shipping labels to the printer in the inventory department.

A second example illustrates why any CTI system in an environment that allows post-call activities should be designed with an understanding of: (1) the entities who will be doing the post-call activities, and (2) their attributes, roles, responsibilities and accessibility while doing them. In this example, the technology being implemented in a particular call center could not route calls to agents doing research unless the agents reverted to a ready status to take calls. This would have been an easy problem to solve if the agents had an "overflow" status that could automatically switch them between doing research at their workstations and answering inbound calls when necessitated by call volume.

Note that this example is not classified as a "call routing" issue because the focus is on doing post-call activities while being automatically accessible in the event of high call volume.

4.5   Technology/Historical Example

Technology/historical examples typically involve one or more databases. For example, major user interface problems can occur when detailed historical records are accessible only through a legacy application that is not integrated into the CTI system. This is particularly true when human agents need information from the database during a call.

4.6   Request/Pre-call Example

Request/pre-call problems can often be resolved during design by determining: 1) what information the caller needs available prior to making the call; and 2) how easy it is for the caller to obtain that information.

For example, some callers have had problems determining their account number when looking at their telephone bills. The printed format is such that the telephone number plus some extra digits is not easily recognized as the account number.

4.7   Request/Call-Routing Example

Request/call-routing issues can occur when call routing is done as part of the task interactions for a request. For example, a CTI system for calls to a financial institution may use the type of request, such as selling and buying of stock, to route calls only to agents who have been certified for that particular type of transaction.

4.8   Request/Task-Interactions Example

As a request/task-interactions example, consider the task interactions involved in a relatively simple request, that of leaving a voice message for someone. The simplest user interface requires only dialing the telephone number and speaking the message after the tone. On the other hand, a complex user interface may require several steps: 1) dialing the telephone number; 2) entering the recipient's mailbox number; 3) choosing the type of message to leave; 4) speaking the message; 5) indicating no special instructions (e.g., "urgent"); 6) indicating whether or not a receipt is required; and 7) entering the fax number for the receipt.

4.9   Request/Post-call Example

A major request/post-call issue is whether or not to design automated workflow for each type of request. For example, an employee self-service system might allow changes to beneficiaries for retirement accounts. If a beneficiary change requires follow-up paper forms containing signatures, then those forms could be identified in the CTI business rules. Thus for the request "retirement-beneficiary changes", the technology/post-call cell would indicate that the system automatically: 1) generates one or more online check lists, and 2) sends appropriate alerts as e-mail.

4.10   Request/Historical Example

Historical information about the types of request can provide very valuable information. For example, an analysis of requests coming into customer service might indicate that one product is generating a high proportion of the calls. Analyzing the reasons for these calls could yield a product fix that would 1) decrease the cost of supporting the current product, and 2) suggest how to create an improved version of the product that would increase market share.

4.11   Caller/Pre-call Examples

As caller/pre-call example, consider a caller population that includes numerous elderly people unfamiliar with automated systems and numerous other callers experienced with many types of communication technology. In this situation the user interface should be designed to deal with caller complexity, such as containing both the typical prompts for callers with rotary telephones and interruptable prompts for experienced callers. In addition, the CTI system might be expanded to include additional modes of access such as the Internet.

4.12   Caller/Call-Routing Examples

As a caller/call-routing example, some systems immediately transfer callers to a human agent if the callers do not enter any information. In these circumstances, skill-based routing (e.g., routing calls to human agents who speak Spanish) involves call-routing based on one or more attributes of the individual caller, rather than on the attributes of the caller's request (see 4.7).

4.13   Caller/Task-Interactions Example

A person's cognitive capacity can be decreased by stress, fatigue or fear. Thus a user interface for callers experiencing strong negative emotions should focus on simplicity rather than additional features. For example, a doctor's office installed a CTI system to force patients to leave voice mail rather than speak to a receptionist. The doctor is in a speciality field in which patients are inherently under a lot of stress. One caller even burst into tears because of the additional stress experienced using the complex system, and consequently is no longer a patient of that doctor. As this example indicates, even expectations about the highly subjective emotional states of callers should, when appropriate, affect CTI design and implementation.

4.14   Caller/Post-call Example

Caller/post-call issues can easily occur in CTI systems that use data from databases controlled by legacy systems. In one such system caller complexity created by changes in marital status caused dramatic negative consequences because of legacy-system post-call activities. The legacy system included the name and address of a caller's divorced spouse in mailing information for newly purchased financial products. As a result of this, the caller's new financial information was mailed to the address of the divorced spouse.

4.15   Caller/Historical Example

A key caller/historical issue in some CTI systems is to be able to search on the caller's name. In some systems, the caller's name, when it is not the same as the customer's name on an account, can only be entered in a comment field that is not indexed for searching. Although in some situations it is not necessary to distinguish between caller and account or customer name, the decision to not be able to search on a caller's name should be a conscious choice based on an understanding of any loss of information or risks involved.

5.   BLENDING OUTBOUND CALLS INTO CTI

As noted in the introduction, "caller" is not an appropriate term when the method of contact is something other than a voice telephone call. The terminology gets even more confused and ambiguous when companies blend outbound calls into the call mix. For example, some airlines have improved the productivity of their customer-service representatives by using automated ("predictive dialer") outbound calls to warn customers of flight delays because of weather. The call is transferred to a human agent only after the call is answered by a human voice, which saves many seconds of a representative's time for each call.

When adding outbound calls to the call mix, it is useful to add a row for "Outbound" to the matrix in Table 1 as a source of complexity. For a sales campaign, the row would contain examples such as those shown in Table 2.

Table 2.   Cell names and examples for an "Outbound" row added to Table 1.

Cell Name	Example
Outbound/Pre-Call	Set attribute thresholds that determine who gets called
Outbound/Call Routing	Set variables that determine which agent gets the screen pop if using skill-based routing for outbound calls
Outbound/Task Interactions	Decisions by the human agents that determine additional specials or offers described to customers based on their comments during the dialogue
Outbound/Post-Call Activities	Order fulfillment for the sales
Outbound/Historical Analysis	Refining the database for future outbound campaigns

6. CONCLUSION

The matrix framework relating "call phase" to "complexity" was designed to aid in design of CTI user interfaces and to help identify potential problems that can decrease caller and/or user satisfaction. The importance of specific issues depends upon the particular CTI system and its context. A short list of general issues identified by use of the matrix includes:

♦   Poor business-process definitions

♦   Inconsistencies in the conceptual or logical models for the CTI system

♦   User interfaces that did not consider caller complexity

♦   Needed information unavailable to agents during task interactions

♦   Lack of integration (e.g., problems with call transfers)

♦   Failure to plan workflow for increased volume due to CTI success

♦   Unplanned consequences of changes

Identifying and solving any one of these problems during design rather than after implementation can easily repay any cost of human factors efforts. It is always more expensive to have to do rework after implementation than to do it right during design. Sometimes it is even impossible to achieve a good fix after implementation because the rework would involve things that can no longer be changed.

14 THE FUTURE OF VOICE INTERACTIVE APPLICATIONS

Daryle Gardner-Bonneau

Michigan State University/Kalamazoo Center for Medical Studies

Key words: Automatic Translation, Augmentative and Alternative Communication (AAC), System Integration, Portability, Medical Transcription

Abstract: *The purpose of this chapter is to explore some of the current issues facing speech technologies as a whole, including the internationalization of applications, portability of technologies, and application and system integration. Usability issues for speech recognition on the desktop are also considered.*

1. THE LARGER PICTURE

It should humble us all to note that, despite the many wonderful developments occurring with respect to speech technology and its application, half the world's population has never encountered a voice interactive application. Nor are they apt to encounter one in their lifetimes. Many have predicted that the world will become more and more dichotomized, based on access to information. Those with the information will have the power; those without it will not. Most of the applications in this book have not reached countries in the Third World to any great extent, and the drivers for many of the existing applications are business interests. One of the exceptions is assistive technology in the area of augmentative and alternative communication (AAC) (see Chapter 6). Thanks, in large part, to the International Society for Augmentative and Alternative Communication (ISAAC), countries all over the world have been introduced to AAC technologies and the promise they hold for bettering the lives of non-speaking people and those with speech impairments. Such efforts are not all that common.

One of the major challenges for human factors professionals in the speech technologies field, as it is throughout the software and telecommunications industries, is the internationalization of applications. Technologically, we have the ability to contact people all over the world by telephone, but communicating with each other around the world can be another matter entirely. I relearned this old lesson when, during the final days of working on this book, I needed to reach, by telephone, one of the book's authors living in France. I called and encountered someone in his office who only spoke French, and that was that. Technology or not, I was not getting anywhere in the conversation!

Europeans have been working on the problem of automatic translation for some time, because the need for this capability is so strikingly clear there. Americans, on the other hand, have been extremely lucky to find business associates throughout the world who speak English. Internationalization of applications, however, will require an understanding of other languages, as well as other cultures, and we can expect that work in this area will escalate over the next decade. It would be nice if automatic translation were merely a matter of directly translating words from one language to another, but it's far from that simple. We use many idioms in our language, and some of them don't translate directly from one language to another. This is just one of many challenges that must be managed. Some years ago, one of the early attempts to assist deaf clients in communicating over the phone was a gesture-to-speech system based on recognition of American Sign Language (ASL). One of the first things that investigators learned was that a direct translation made no sense at all. In retrospect, it became apparent that ASL has its own grammar and structure that are quite efficient, but also quite different from American English. While the problem of translating written or spoken languages may be less severe, analogous problems exist. In addition, if speech applications involve associated visual displays, their user interfaces must be translated, as well, and this task poses its own problems for many applications. Icon design for graphical user interfaces (GUIs) serves as one example. Symbols that are recognizable and acceptable to English-speaking audiences are sometimes neither recognizable nor acceptable, and may even be offensive, to people from other cultures. Designing a symbol that conveys the same meaning to people all over the world, without being offensive to anyone, can be a daunting task. No doubt, such work will keep many human factors professionals busy during the years to come.

2. PORTABILITY AND SYSTEM INTEGRATION

Portability and system integration are two of the biggest trends in computing. We want to work from anywhere and have everything we need at our fingertips. In the U. S., laptops are becoming an essential business tool for work "on the road." In Europe, cellular telephones are even more popular than they are in the U.S. In addition, there is consumer pressure for more and more functionality to be added to these devices. The extent to which speech technologies will be incorporated within laptops and cellular telephones remains to be seen.

At this point, speech-enabling a laptop computer is a risky proposition, unless you happen to have a laptop that is certified, specifically, by the developer of the speech recognition tool you wish to buy. Speech technology developers make no guarantees their software will work with hardware other than that which they certify. Unfortunately, these developers certify only a small fraction of the laptops available and aren't in a position to be able to test their products with all of the hardware that potential purchasers might be using. This is a significant problem for those who wish to use speech on a laptop. Additionally, some laptops are too noisy to support speech recognition, and the noise rating is a feature of laptops that many potential users ignore when they make a laptop purchase. Furthermore, information about a laptop's noise rating is not readily available, and requires some digging on the part of the consumer. Finally, using speech recognition today requires a fairly powerful computer system, if you want to control your software and engage in voice dictation at more than a snail's pace. Not all of these problems are human factors issues, but all of them relate to the issue of portability, and to the challenges faced by technology developers in meeting this goal.

Just as serious are the challenges posed by system integration requirements, both from the user point of view and the technical point of view. Both Martha Lindeman, in her chapter on Computer-Telephony Integration (CTI), and Harry Blanchard and Steve Lewis, in their chapter on voice mail, described some of the human factors issues involved in integrating multiple, related services in such a way that they work together effectively. Another type of integration problem occurs when a custom speech application is being integrated into a larger, complex system. An example from medicine will serve to illustrate.

One of the major potential markets for ASR dictation systems is in the world of medicine. Millions of dollars are spent each year on medical transcription services that translate a physician's handwritten notes or notes dictated into a tape recorder into a permanent computerized record. If there was ever the case for "eliminating the middle man," this is it. In Chapter 1, John Karat and his colleagues described the human factors work that went into the development of MedSpeak, a speech dictation system for generating radiology reports. One of the reasons MedSpeak has enjoyed success is that there was involvement from potential users in its development from the very beginning. This helped to ensure that the resulting system would be compatible with physician models of the task and could be seamlessly integrated into their immediate work environment with a minimum of disruption. In and of itself, this was no small feat, because physicians are highly skeptical of new technology, have very little time to devote to development efforts, and are known to be intolerant of systems that don't work. Although MedSpeak, and other systems like it, have been developed for several niche areas within medicine, there is another level of integration that must be achieved before these systems will be commonly accepted in the medical arena. They must be integrated within the hospital environment, as a whole, and integrated with other hospital computer systems that must access their data. This can be a major headache for vendors, because there are often multiple hospital software systems and, frequently, at least one of them is a legacy system (often, for billing) that was written in, say, COBOL. Yet, this level of integration must be achieved if speech technologies are going to change the way that hospitals conduct their business.

As to integration within the hospital environment, IBM was smart to choose an application domain in which there was a reasonable chance of success. Radiology is a clearly distinct department in hospitals. Radiologists conduct their work in a confined area, and have a quiet place available to them when they dictate reports. Though information must be received from and passed to other points in the hospital, for the most part, the radiologists do their work at a single location. Such is not the case at all for many other types of physicians and, thus, there would be more barriers to system integration. For example, integrating a dictation application for, say, internal medicine specialists, would be much more difficult. They move frequently from place to place within the hospital, and they interact with many other people while doing their work. For speech to work for them might require, for example, the redesign of some of their work areas to to support dictation stations.

3.   DESKTOP SPEECH RECOGNITION

Despite major technology breakthroughs in large vocabulary, speaker-dependent recognition over the past two years, and the mass marketing of shrink-wrapped dictation software, speech recognition has yet to become a productivity tool for the masses. Many of my colleagues work in the computer industry as user interface designers and usability specialists. Recently, I conducted an informal poll to get some idea as to the extent to which this technology is being used. The results were rather disappointing. Very few people were using this technology at work. The reasons for this are many, but among the most common is that people who "live" in cubicles can't disturb others by using a speech recognition package to do their work. But more than that, many people still hold that the view that the technology is not yet usable; it's buggy and it's not well integrated within the desktop environment. They are probably right.

The problem of mode switching between dictation and command and control still plagues us, despite several years of work on the problem. It is extremely irritating for users to issue a system command and have it appear as dictated text, or worse yet for the computer to execute a command (the user knows not what) in response to dictated text. There's no question that the most recent software packages have improved in this regard, but there is still more work to be done. In addition, processing of dictation, while very fast when users are running only one application on a powerful machine, can slow to a crawl when users open several applications at one time. Unfortunately, for many users, opening several applications at once is the typical mode of operation. If user group conversations on the Internet are any indication, various problems occur when users attempt to control applications by voice, even when the vendors specifically indicate that a particular piece of software can be run using their speech system. Users find that some actions are no longer possible, or must be accomplished in other, less efficient ways than they otherwise would if using keyboard command and control. A colleague of mine was recently involved with usability testing of a number of dictation products currently on the market. His conclusion: "The babies are still very ugly."

In addition, there exists the problem created by the fact that speech recognition systems are not isolated, standalone systems, as some might like to believe. The performance of these systems is, as everyone already knows, affected by changes in the voice of the user (e.g., getting a cold can create tremendous problems when using these systems), noise in the environment, and a number of other factors over which the system developer may have very little control. However, it also appears that speech recognition system performance depends, heavily, on the particular microphones and sound cards that are being used. Users see dramatic differences in performance from one microphone to another and from one sound card to another. In addition, though some developers deny it, there seems to be an interaction between sound cards and microphones that plays a role in system performance. Thus, in many ways the speech recognition developers don't control their own fates, unless they're also looking seriously at microphone and sound card issues. Although they can designate specific sound cards and microphones for use with their software, they cannot stop purchasers from using what they have which, in many cases, will not be what has been recommended. When performance is poor, it is not the microphones and the sound cards that will be blamed, it is the speech recognition software. Thus, robust recognition, in light of all these elements, remains a goal for speech recognition systems.

4. PROSODY IN SPEECH SYNTHESIS

As Francis and Nusbaum indicated in Chapter 3, the primary indicators of success for synthetic speech systems are intelligibility and naturalness. Though still an issue under certain conditions of use, intelligibility is a goal that has been met for quite some time. Naturalness, on the other hand, still poses some challenges. Although systems around the world are getting better and better in this regard, conveying emotions – along with appropriate inflection and timing of utterances – will remain important to the success of these systems in practice.

5. PARTING COMMENTS

This chapter has described some of the current issues in the application of speech technology. Many of these issues have human factors components and will be resolved through the use of human factors methods, tools, and techniques. Some of those resolutions are just around the corner; some are years of work away. Clearly, we can learn much by involving real users in the design process. We should listen to what they say. But more importantly, we need to watch what they do – on their own turf and in their own working environments – if we want to solve their problems and bring usable speech technology applications to the masses.

INDEX